Biointensive
Integrated Pest Management
for Horticultural Crops

Biointensive Integrated Pest Management for Horticultural Crops

Chief Editor

Anil Kumar

Assistant Professor-cum-Scientist
Department of Entomology
Sugarcane Research Institute
Dr. Rajendra Prasad Central Agricultural University
Pusa-848125, Samastipur, Bihar, India

Associate Editors

Swati Saha

Scientist
ICAR-IARI Regional Station
Pune, Maharashtra, India

Jaipal Singh Choudhary

Scientist
ICAR-Research Complex for Eastern Region
Research Centre, Ranchi, Jharkhand, India

CRC Press is an imprint of the
Taylor & Francis Group, an **informa** business

NEW INDIA PUBLISHING AGENCY
New Delhi – 110 034

First published 2022
by CRC Press
4 Park Square, Milton Park, Abingdon, Oxon, OX14 4RN

and by CRC Press
6000 Broken Sound Parkway NW, Suite 300, Boca Raton, FL 33487-2742

CRC Press is an imprint of Taylor & Francis Group, an Informa business

British Library Cataloguing-in-Publication Data
A catalogue record for this book is available from the British Library

Library of Congress Cataloging-in-Publication Data
A catalog record has been requested

ISBN: 978-1-032-18908-6 (hbk)
ISBN: 978-1-003-25692-2 (ebk)

DOI: 10.1201/9781003256922

Foreword

Horticulture sector includes a wide range of crops namely fruit crops, vegetables crops, ornamental crops, medicinal and aromatic crops. It has improved economic status of farmers and increases the nutrient availability of human being. The horticultural crop production has become a key driver for economic development in many of the states in the country and it contributes 30.4 per cent to GDP of agriculture. Globally, India is the second largest producer of fruits and vegetables. Though, India achieved self-sufficiency in food grain production, but still the country has not achieved self-sufficiency in production of horticultural crops. The challenge before the crop protection scientist is to increase yields from the existing land without harming the environment and resource base and to reduce losses caused by pests. The future strategies should, therefore, be carefully planned to sustain production with least disruption to the agro ecosystem. There has been an increasing demand to not only produce larger quantities, but also towards sustainable agriculture where products are simultaneously environmental friendly, socially fair and economically beneficial. This can be achieved by adopting eco-friendly (Biointensive Integrated Pest management) strategy. It includes ecological and economic development into agricultural design and decision making system and also addressing public concern about environmental quality and food safety. Biointensive Integrated Pest management strategy begins with steps to accurately diagnose the nature and source of pest problems, and then relies on a range of preventive tactics and biological controls to keep pest populations within acceptable limits. Reduced-risk pesticides are used if other tactics have not been adequately effective, as a last resort, and with care to minimize risks. In this context, the present book **"Biointensive Integrated Pest Management for Horticultural Crops"** edited by Dr. Anil Kumar as Chief Editor and Swati Saha, and Jaipal Singh Choudhary as Associate Editors, will be useful to teachers, researchers, students and extension workers.

I hope this publication will be of great value to those who are engaged in growing horticultural crops. I appreciate the meticulous efforts of editors in bringing out such an useful publication.

Date: 12.12.19

(S.P. Singh)
Professor & Head
Department of Entomology
Dr. Rajendra Prasad Central Agricultural University
Pusa-848 125, Samastipur, Bihar

Preface

In fact insect pest problem rapidly increases in current scenario due to changing climatic condition, indiscriminate use of insecticides which leads to decrease biodiversity of natural enemies and effect on human being. There is a need of hour to create awareness for promoting environmentally sustainable horticultural production.

The book, **Biointensive Integrated Pest Management for Horticultural Crops** has been planned with chapters and the scope of each chapter has been specified by the author. The book deals with the most recent biointensive integrated approaches utilizing components such as bioagents, botanicals and microbial pesticides, physical and mechanical methods, cultural methods such as crop rotation, summer ploughing, intercropping, pruning, mulching, spacing, planting date, trap cropping, etc., biorational chemicals and tolerant cultivars. It is our great pleasure to express our sincere thanks to the publisher, New India Publishing Agency for producing this book in a systemic way with quality within a period of time.

It is earnestly hoped that the book will be a useful reading to all those who are interested in Biointensive Integrated Pest Management for Horticultural Crops.

We request all the readers for rendering valuable suggestions for future improvement of this edition.

Date: 12.12.2019

Anil Kumar
Chief Editor

Contents

Vegetable Crops

Flower Crops

Medicinal, Aromatic and Spice Crops

List of Contributors

Abbas Ahmad, Department of Entomology, Dr. Rajendra Prasad Central Agricultural University Pusa-848125, Samastipur, Bihar

Anil Kumar, Assistant Professor-cum-Scientist, Department of Entomology, Sugarcane Research Institute, Dr. Rajendra Prasad Central Agricultural University, Samastipur-848125, Bihar

Ayan Das, Research Scholar, Department of Agricultural Entomology, Bidhan Chandr Krishi Viswavidyalaya, Mohanpur, West Bengal-741252

B. S. Gotyal, Senior Scientist, ICAR-CRIJAF Barrackpore, Kolkata-700121, West Bengal

Beer Bahadur Singh, Assiatant professor-cum-Jr. Scientist, Nalanda College of Horticulture Noorsarai, Nalanda, Bihar

Bikash Das, Principal Scientist, ICAR, Research Complex for Eastern Region Research Centre Ranchi-834010, Jharkhand, India

Davendra Kumar, ICAR-Indian Agricultural, Research Institute, New Delhi-110012

Fouzia Bari, Department of Entomology UBKV, Cooch Behar, West Bengal

Girish K.S., Scientist (Agriculture Entomology) ICAR-Directorate of Floricultural Research Pune, Maharashtra

Gundappa Baradevanal, ICAR-Central Institute for Subtropical Horticulture, Rehmankhera Lucknow-226101, Uttar Pradesh

Jaipal Singh Choudhary, Scientist, ICAR, Research Complex for Eastern Region Research Centre Ranchi-834010 Jharkhand, India

Kavya shree, Assistant Professor, Department of Horticulture, Dr. Rajendra Prasad Central Agricultural University, Pusa Samastipur, Bihar

Kuldeep Srivastav, Principal Scientist, ICAR, National Research Centre on Litchi, Muzaffarpur-842002, Bihar

Kumarnag K.M., ICAR- Indian Agricultural, Research Institute, New Delhi-110012

Lalita Rana, Assistant Professor, Department of Agronomy, Dr. Rajendra Prasad Central Agriculture University, Pusa 848125, Smastipur, Bihar.

M. Raghuraman, Department of Entomology and Agril. Zoology, Institute of Agricultural Sciences, Banaras Hindu University, Varanasi-221002, Uttar Pradesh

Mahalle, R.M., Department of Entomology and Agricultural Zoology I.Ag.Sc. BHU, Varanasi Uttar Pradesh

Manish Chandra Mehta, Department of Entomology and Agril. Zoology, Institute of Agricultural Sciences, Banaras Hindu University, Varanasi-221 005, Uttar Pradesh

Nagendra Kumar, Assistant Professor-cum-Scientist Department of Entomology, Dr. Rajendra Prasad Central Agricultural University, Pusa-848125, Samastipur, Bihar

Ponnusamy, N., Department of Entomology, Dr. Rajendra Prasad Central Agricultural University, Pusa-848125, Samastipur, Bihar

Priyanka Kumawat, Ph.D.Scholar, Department of Horticulture, Swami Keshwanand Rajasthan Agriculture University, Bikaner, Rajasthan

R. Maurya, Department of Horticulture, I.Ag.Sc., BHU, Varanasi, Uttar Pradesh.

R.N. Singh, Department of Entomology and Agricultural Zoology, I.Ag.Sc. Banaras Hindu University, Varanasi, Uttar Pradesh-221005

R.R Dhole, Department of Entomology and Agricultural Zoology I.Ag. Sc. Banaras Hindu University, Varanasi, Uttar Pradesh-221005

S. Sahoo, Associate Professor, Department of Entomology, Dr. Rajendra Prasad Central Agricultural University, Pusa-848125 Samastipur, Bihar, India

S. Satpathy, ICAR-CRIJAF, Barrackpore, Kolkata-700121, West Bangal

S.K. Senapati, Professor, Department of Entomology College of Agriculture, Uttar Banga Krishi Vishwavidyalaya Majhian, Dakshin Dinjapur-733133, West Bengal

S.P. Patil, Department of Agril.Entomology MPKV, Rahuri Dist. Ahmednagar India

Sagar Tamang, Purba medinipur KVK, Mulakhop Nandakumar, 721632

Sandeep Kumar, Department of Fruit Science, College of Horticulture and Forestry, Jhalrapatan Campus, Rajasthan

Sapna Panwar, Division of Floriculture and Landscaping, ICAR-Indian Agricultural Research Institute, New Delhi-110012

Sudarshan Chakraborti, Professor, Department of Agricultural Entomology, Banaras Hindu University, Mohanpur, West Bengal 741252

Suprakash Pal, Department of Agricultural Entomology, Uttar Banga Krishi Viswavidyalaya Cooch Behar, West Bengal-736165

Swati Saha, Scientist, ICAR-IARI, Regional Station, Pune, Maharashtra

Tanweer Alam, Department of Entomology, TCA Dholi, RPCAU, PUSA

Tarak Nath Saha, Senior Scientist, ICAR-DFR, Pune, Maharashtra

Udit Kumar, Assistant Professor, Department of Horticulture, Dr. Rajendra Prasad Central Agriculture University, Pusa, Smastipur, Bihar

Umesh Das, Department of Agril. Entomology Uttar Banga Krishi Viswavidyalaya, Cooch Behar, West Bengal-736165

V. Ramesh Babu, ICAR-CRIJAF, Barrackpore, Kolkata-700121, West Bengal

Introduction

1

Scope and Importance of Horticultural Crops

Swati Saha, Udit Kumar, Tarak Nath Saha and Kavya Shree

Introduction

Horticulture is the science and art of propagation, production, utilization, improvement, harvesting and marketing of crops such as fruits, vegetables, ornamental plants, spices & condiments, medicinal & aromatic plants, plantation crops, postharvest technology etc. The term is derived from the Latin word *hortus* mean garden and *cultura* means to cultivate. These crops are diverse which includes annuals to perennials, edible to ornamentals, temperate to tropical, humid to desert, etc. Horticultural plants help to sustain and enrich people by providing nutritious food, medicines, enhances the beauty of surroundings, besides reducing carbon footprint. They also serves as raw material for various industries, such as processing, pharmaceutical, perfumery and cosmetics, chemical, confectionery, oils and paints, etc.

Importance of Horticultural Crops

The horticulture crops are a source of variability in farm produce as well as human diets. The comparative production per unit area of horticultural crops is much higher than field crops. They are not only a source of nutrients, vitamins, minerals, flavour, aroma and dietary fibres, but also contains health benefiting compounds and medicines. Ornamental horticulture have aesthetic value and protects the environment. These crops are also useful for cultivation in wasteland, poor quality soil, slopy, uneven or undulating lands. Mango and cashew nut are cultivated on a large scale in hilly and hill back areas. Such crops are of high value, labour intensive and generate employment throughout the year. Further they have national and international demand and are a good source of foreign exchange.

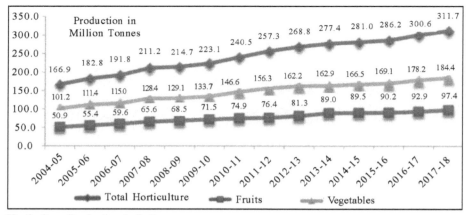

Horticulture Production in India
Source: Ministry of Agriculture, GOI

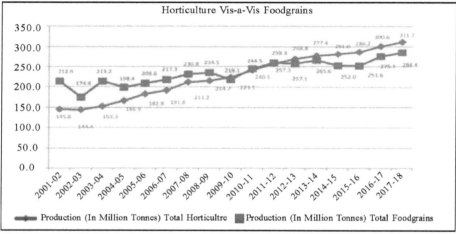

Horticulture *vs* food grain production over the years
Source: Ministry of Agriculture, GOI

Prospects of horticultural crops in India

Horticulture sector is recognised to have the potential to raise the farm income, provide livelihood security and earn foreign exchange. As per the latest estimate (2018-19) by NHB, the total horticulture production in the country is 314.67 million tonnes from an area of 25.87 million hectares. India is the second largest producer of fruits and vegetables in the world after China. Horticultural crops constitute a significant portion of the total agricultural produce in India. They cover a wide cultivation area and contribute about 28 per cent of the Gross Domestic Product (GDP).

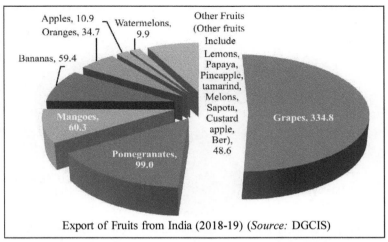

Export of Fruits from India (2018-19) (*Source:* DGCIS)

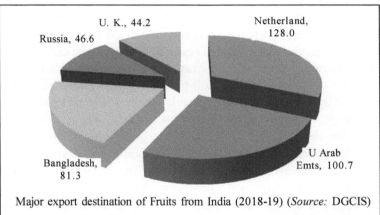

Major export destination of Fruits from India (2018-19) (*Source:* DGCIS)

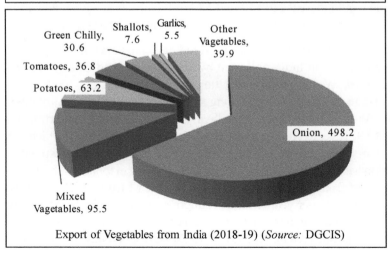

Export of Vegetables from India (2018-19) (*Source:* DGCIS)

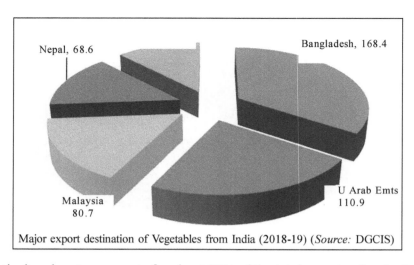

Major export destination of Vegetables from India (2018-19) (*Source:* DGCIS)

Horticultural sector accounts for about 37% of the total exports of agricultural commodities, and the exports have recorded sustained rising trend. This sector is improving the productivity through research and development, enhancing the share of value added products, geographical diversification of exports and enhancing the infrastructure including cold storage and rural roads. The government of India has proposed to double farmer's income by the year 2022. It is increasingly being recognised that horticulture will remain an integral component for the strategy to achieve this goal. The budgetary share of consumers for fruits and vegetables in total food expenditure has increased across all income groups (Kumar *et al.*, 2011). The export of the horticultural commodities has also shown greater prospects and this trend is likely to continue. In this context, it is worthwhile to note that by 2030 the demand for fruits and vegetable would increase to 110 and 180 million tonnes, envisaging an increase of 155 and 95%, respectively over the base year of 2000 (ICAR 2011).

In fruit production, Andhra Pradesh (excluding Telangana) emerged as the largest producer of fruits in India, followed by Maharashtra and Uttar Pradesh. While in vegetables, West Bengal and Uttar Pradesh turned out to be the biggest producers. In floriculture the leading states are Tamil Nadu, Karnataka, West Bengal, Andhra Pradesh and Madhya Pradesh. A study undertaken by Jha *et al.* (2019) revealed higher growth in area under vegetables, which increased from 6.1 m ha in 2001-02 to 10.3 million ha in 2016-17. Correspondingly the production has increased from 88.6 m tonnes to 175.0 m tonnes with a productivity improvement from 14.4 t/ha to 17.0 t/ha. In case of fruits, the area growth was at the rate of 3.5% per year, with a production growth of 5.6%. The growth rate of horticultural commodities including cash crops, spices and fruits and vegetables accounts for about 1.98%, 2.33% and 19.24%, respectively. During 2001-02 to

2016-17, the total area under the horticultural crops increased from about 16.5 million ha to 25 million ha, at an annual trend growth rate of about 3.0%. Correspondingly total production of horticulture commodities increased from 146 million tonnes to 295 million tonnes at an annual growth rate of 5.8%. The growth in the productivity of the horticultural sector along with the area and price changes has led to an improvement in the share of horticultural commodities in the value of output. Mission for Integrated Development of Horticulture (MIDH) implemented by Government of India by adopting an end to end approach for increasing production of horticulture crops and reducing post-harvest losses is an important step for the development of horticultural sector.

A viable option for diversification

Day by day the land area availability for agriculture is shrinking due to expanding urbanization. But at the same time demand for higher productivity and returns from the cultivable land is increasing rapidly. This leads to gradual change in the cropping pattern, land use system, input utilization, marketing and above all the monetary returns in the Indian agriculture. Therefore, the quest for higher productivity is leading to serious problems of soil salinity, alkalinity, soil degradation, depletion of ground water table, etc. These factors provided an ideal condition for shifting towards diversification, mostly in favor of horticultural crops such as fruits, vegetables, spices, plantation crops and ornament crops. Horticulture crops specially the fruit crops are more resilient to change in weather conditions. The vegetables augment the income of small and marginal farmers. The vegetable cultivation is remunerative and its cultivation absorbs large scale man power. Due to labour intensive nature, an increased area under the crop would also provide employment opportunity to youth in agriculture. In that context, the diversification towards high value commodities is labour absorbing and contributes to employment generation (Joshi *et al.,* 2004). Water utilisation is very low, minimising the risk of crop failure and it can be done on smaller farms. Multiple crops are planted simultaneously to get more yield and to use the maximum of the fertilisers. This sector enables the population to eat a diverse and balanced diet for a healthy lifestyle. Thus it became a key driver for economic development in many of the states in the country where Division of Horticulture of Indian Council of Agricultural Research is playing a pivotal role. The need for meeting the minimum nutritional level of the diet of a common man is assuming greater significance today. Horticultural crops i.e. fruits and vegetable acquire a place of important as protective food. They provide much needed health supporting vitamins, minerals. Thus, diversification towards the horticultural commodities is seen not only in value terms, but in terms of area as well.

Commercial Horticulture

Commercial Horticulture involves commercial activity for the production and management of fruits, vegetables, ornamental plants, nursery plants (both in open or protected structures); hi-tech nurseries; grading, harvesting and processing of the horticultural produce etc. It generally involves high value and low volume crops. Now-a-days commercial horticulture is producing high quality traditional and exotic fruits, vegetables, flowers, etc. Among the horticulture sector, floriculture is growing at the rate of 12-15 % per annum. Around 500 ha is under protected cultivation of floricultural crops. The main export components are dry flowers, fresh flowers and potted ornamentals. Employment opportunities provided by this sector to the farm population are production, transportation, processing and marketing operations, in addition to the entrepreneurs seeking self employment. Keeping in view its importance much emphasis has been laid to augment the production of horticultural crops in our national plans. Horticulture crops perform a vital role in the Indian economy by generating employment, providing raw material to various food processing industries, and higher farm profitability due to higher production and export earnings from foreign exchange.

The area and production of various horticultural crops are as follows (NHB 2018-19).

S.No.	Particulars	Area ('000 Ha)	Production ('000 MT)
1.	Fruits	6530	96754
2.	Vegetables	10436	187474
3.	Flowers	339	2858
4.	Medicinal and Aromatics	717	889
5.	Plantation crops	3762	17991
6.	Spices	4086	8590
7.	Honey	-	113
	Total	**25870**	**314671**

Employment opportunities in horticulture

Horticulture is growing field of agriculture that deals with the cultivation and processing of plants, fruits, vegetables, flowers, aromatic and medicinal plants. People attraction towards horticulture sector increases because government focusing on horticultural sector due to its significant contribution in the economy as employment generators and industrial raw material provider. Liberalization, privatization and globalization of world trade also generated the opportunities of employment for youths across the country. Horticulture is an extremely diversified field with unlimited career opportunities in a variety of areas (www.khake.com). Horticulture education is emerging field of agriculture that providing immense employment opportunities to rural and urban youths across the country.

A horticultural graduate can get job on different posts in academics, research technician, consultancy, banking, in governmental institutes, private firms cooperative and NGOs. Thus, a horticultural graduate can get employment on various posts such as horticulturist, floriculturist, olericulturist, pomologist technician, horticultural consultant, horticultural therapist, botanical garden worker, green house manager, landscape architect, viticulturist, nursery manager and horticultural marketers in governmental and private institutions along with an alternative of self-employment through entrepreneurship development (Bairwa *et al.*, 2014). Horticulture has become one of the best career options due to ample employment potential from academic to research and development in private as well as government sector in India as well as abroad. Different agricultural universities also employ horticultural postgraduates for distinct posts from concerned field of their specialization. Buxton and Cassady (2011) Bairwa *et al.* (2014) and Meena (2015) identified different major horticultural job positions in various departments, institutes and privates firms for horticultural graduates, post graduates and doctorates degree holders *viz.* Floriculturist, Olericulturist, Pomologist, Botanical / Horticultural Garden Worker, Horticultural Consultant, Therapist, Greenhouse Manager, Landscape Architect, Nursery Manager, Horticultural Marketer, Viticulturist etc.

Economic Importance

Horticulture forms an integral and important component in the economy of a nation. Horticultural crops constitute a significant segment of the total agricultural production of a country. The importance of horticulture can be substantiated by its benefits like high export value, high yield per unit area, high returns per unit area, best utilization of wasteland, provision of raw materials for industries, whole engagement by a grower/laborer, production of more food energy per unit area than that of field crops, better use of undulating lands, and stabilization of women's empowerment by providing employment opportunities through processing, floriculture, seed production, mushroom cultivation, nursery preparation, etc. In addition, fruits and vegetables constitute the important energy-giving material to the human body. It also improves the economic condition of many farmers, and it has become a means of improving livelihood for many unprivileged classes too. Flower harvesting, nursery maintenance, hybrid seed production and tissue culture, propagation of fruits and flowers, and food processing are highly specialized remunerative employment options.

Scope of Horticulture in India

India is a vast country with variety of climate and edaphic conditions which can be successfully exploited by growing horticultural crops.

- Climates are varying from tropical, subtropical and temperate regions besides humid, semi-arid, arid in which a number of horticulture plants can be grown throughout the year.

- Availability of wide range of soils like loamy, alluvial, laterite, medium black, rocky shallow heavy black sandy etc. From this, large crop areas under horticulture can be grown with very high level of adaptability.

- To meet the nutritional requirements in terms of vitamins and minerals, minimum of 85 g of fruits and 200 g of vegetables per head per day with population of above 125 crore people.

- For providing raw material to small scale industries like silkworm, lack, honey, match, paper, canning, and dehydration etc. horticulture has wide scope.

- The vast track of waste land, problematic soil, desert land can be utilized for hardy fruit crops, medicinal plants and other suitable crops.

- With the development of rapid road and railway network fast development of communication and transport system create wide scope for horticulture development particularly in transporting the perishable commodities and products.

- Horticulture sector contribute immensely by generating skill employment opportunity for labour and human being. This sector generates 860 man days/annum for fruit crops as against 143 man days/annum for cereal crops and the crops like grapes, banana and pineapple demand 1000-2500 man days per annum.

- Horticultural commodities like spices, coffee, tea, etc fetch 10-20 times more foreign exchange per unit weight than cereals. Therefore, taking the advantage of globalization of trade, nearness of big market and the size of production, this sector can provide scope for accelerated growth.

Challenges

- Horticulture sector in India does not enjoy a safety net like the Minimum Support Price (MSP) as in case of foodgrains.

- Low productivity of horticultural produce compared to developed countries.

- Excessive use of pesticides resulted in refusal of export consignment especially in fruits, vegetables and spices.

- Lack of good cold chain storage and rural networks to extend the life of perishable products.

- Very less or limited input by machinery and equipment so it is tough to minimize the time restraints.

- Higher input costs than food grains make it a difficult set up, especially when there is no support from the local governments to the smaller farmers.

- It gets challenging for marginal farmers to cope with the high price fluctuations, both in domestic and export market.

- Limited availability of market intelligence, mainly for exports makes it a tougher option to choose.

Suggestions

- Location specific research and development programme in horticulture.

- Development and standardization of varieties suitable for cultivation in non-traditional areas.

- To adopt technology led development in Horticulture and entrepreneurship development.

- Implementation of the scope of nutrient dynamics and interaction in horticultural crops.

- Promotion of the use of insect pollinators for improving productivity and quality of the crops.

- Adoption of Integrated Pest Management (IPM) strategies for production of horticultural crops.

- Enhancing the postharvest processing, value addition activities and market support in horticulture crops.

- Development of modified atmosphere packaging for long storability & transportation of fruits, vegetables and flowers.

- Need to spread the export destination of horticulture produce.

- Bio-energy and solid waste utilisation to make horticulture more efficient and eco-friendly.

- Development of a comprehensive plan for coordination and monitoring of R&D programmes at national level as well as to serve as knowledge repository in Horticulture sector.

Way Forward

- The diversification in the agricultural sector mainly of the horticulture sector has become a major source of positive growth for the sector itself and for the nation.

- It has emerged as a promising source of income acceleration, employment generation, poverty alleviation and export promotion.
- India can emerge as a bigger producer and exporter if sufficient emphasis is given to resource allocation, infrastructure development, more R&D, technological upgradation and better policy framework for horticulture sector.
- Horticulture sector with strong forward and backward linkages as an organised industry can stimulate and sustain growth.

References

Anonymous (2011). Vision 2030, Indian Council of Agricultural Research, New Delhi.

Anonymous (2019). Horticulture Statistics – Advance Estimates. Horticulture Statistics Division, Department of Agriculture, Cooperation and Farmers Welfare, Ministry of Agriculture and Framers Welfare, New Delhi.

Bairwa, S.L., Kushwaha, S., Meena, L.K., Lakra, K. and Kumar P. (2014). Agribusiness Potential of North Eastern States: A SWOT Analysis. Edited by Singh *et al.*, 2014 "Agribusiness Potentials in India: experience from hill states". EBH Publishers (India) Guwahati, New Delhi. Pp 544-556.

Bairwa, Shoji Lal, Kalia, Abhishek, Meena, L.K., Lakra, Kerobim and Kushwaha, Saket (2014). Agribusiness Management Education: A Review on Employment Opportunities, *International Journal of Scientific and Research Publications*, 4(2): 536-538.

Bairwa, Shoji Lal, Kumari, Meera and Meena, L.K. (2015). Developing mobile based agri retailing (MBAR) model for high value agricultural and livestock products, *International Journal of Agricultural Science and Research* (IJASR), 5(6): 125-130.

Buxton, J. and Cassady, C. (2011). Careers in horticulture and related fields, College of Agriculture, University of Kentucky, UK. FPJ (2015) Green way – Career in horticulture, Free press journal, Daily Newspaper, 6th April, 2015.

Girish K. Jha, A. Suresh, Bhoopesh Punera and P. Supriya (2019). Growth of horticulture sector in India: Trends and prospects, *Indian Journal of Agricultural Sciences,* 89(2): 314-21.

Joshi P.K., Gulati A., Birthal P.S. and Tewari L. (2004). Agriculture diversification in south asia: Patterns, determinants and policy implications. *Economic and Political Weekly* 39(24): 2457-67.

Kumar P., Kumar A., Parappurathu, S. and Raju, S.S. (2011). Estimation of demand elasticity for food commodities in India, *Agricultural Economics Research Review*, 24(1): 1-14.

Kumar, Pravin, Bairwa, Shoji Lal, Madguni, Omprakash and Bairwa, Suresh (2014). Emerging Business Opportunities in Agroforestry, Edited by Rai and Goyal "Agriculture and Rural Development in India". M.R.F. Publication, Varanasi. Pp. 326-328. ISBN 978-81-926935-6-9.

Meena, V.S. (2015). Career Opportunities in horticulture, Employment News, 11th July 17th July 2015, i.e. No. 15 8. Singh, K. P., Nair, Beena and Chand, Prem (2013) Career in Horticulture, *Employment News Weekly*, Vol. XXXVIII NO. 31 Pp, 56, 2-8 November 2013.

Web India (2015). Horticulture: job prospects & career options, http://career.webindia123.com/career/options/basic_environmental_science/horticulture/jobprospects.htm

www.khake.com/page21.html

2

Biointensive Integrated Pest Management

Anil Kumar and S.K. Sahoo

Biointensive IPM incorporates ecological and economic factors into agricultural system design and decision making while addressing public concerns about environmental quality and food safety. It provides suitable guidelines and options for the effective management of pests and beneficial organisms in an ecological context in a holistic self-sustaining manner. The flexibility and environmental compatibility of the Biointensive IPM strategy make it useful in all types of cropping systems particularly in an agro-biodiversity rich ecosystem.

Component of Biointensive IPM

- Proactive strategies
- Reactive strategies

Proactive Strategies

It is also called Cultural Control Methods. It is the manipulations of the agro ecosystem that make the cropping system less friendly to the establishment and proliferation of pest populations. During Proactive Strategies three things are kept in mind as follows:

- Increase of below ground diversity
- Increase of above ground diversity
- Appropriate plant cultivars.

Cultural controls strategies include

1. **Genetic diversity** of a particular crop may be increased by planting more than one cultivar.

2. **Species diversity** of the associated plants can be increased by allowing trees and other native plants to grow in fence rows or along water ways.

3. **Crop rotation i.e.** growing different crops alternatively alters the environment both above and below ground, usually to the disadvantage of pests of the previous crop grown.

4. **Multiple cropping** is the sequential production of more than one crop on the same land in one year. Depending on the type of cropping sequence used, multiple cropping can be useful as a weed control measure, particularly when the second crop is inter planted into the first.

5. **Inter planting** is seeding or planting a crop into a growing stand, favoring the microclimate (e.g., timing, wind protection, and less radical temperature and humidity changes) as well as keeping the soil covered, protect soil against erosion from wind and rain.

6. **Inter cropping** reduces the attraction of pest to the host, adversely modify the microclimate of the pest habitat by growing two or more crops in the same, alternate, or paired rows in the same area.

7. **Strip cropping** is the practice of growing two or more crops in different strips across a field wide enough for independent cultivation. It increases the diversity of a cropping area, favours the crop against its pests and may act as a reservoir and/or food source for beneficial organisms.

8. **Trap crop** is generally used prevents the insect from reaching the main crop.

9. **Selection of seed and cultivar**
   - Use of disease & pest-free seed and nursery stock is important in preventing the introduction of diseases and pests in a new area.
   - Use of resistant varieties released by researchers or simply by collecting non-hybrid seeds from healthy plants in the field.

10. **Clean cultivation:** Destruction of crop residues and removing and destroying the overwintering or breeding sites of the pest will remove residual pest population as well as preventing a new pest from establishing on the farm.

11. **Spacing i.e.** distance between plants and rows, the shape of beds, and the height of plants influences the development of plant diseases, obnoxious pests and weed problems. Proper spacing will provide better air flow and decrease the incidence of plant diseases and population build up by pests.

12. **Altered planting dates** can at times be used to avoid specific insects, weeds, or diseases by depriving the organisms of their required environmental conditions.

13. **Optimum growing conditions** make the plants grow quickly and healthy which can compete with and resist pests better than slow-growing, weak plants.

14. **Mulches** are useful for suppression of weeds, insect pests, and some plant diseases.

15. **Biotech Crops:** Gene transfer technology is being used by several companies to develop cultivars resistant to insects, diseases, and herbicides.

16. **Water management**
   - Flood irrigation help to reduce incidence of termite
   - Close irrigation reduce incidence of ESB of sugarcane

17. **Nutrient management**
   - Balanced dose of nutrients should be applied for growing crops
   - Excess nitrogen lead to increased incidence of BPH and gall midge

18. **Pruning of crop**
   - Some pests are reduced by clipping of apical portion of plants and old infested leaves

Reactive strategies

There are three methods employed for control of insect pests through biological control:

1. Introduction or Importation or classical biological control

This consists of foreign exploration for exotic natural enemies, their importation and their release to control of pest that has been accidentally introduced. So it necessitates importation and establishment of non native natural enemy population for suppression of native or non-native organism. Foreign exploration is conducted to identify and collect natural enemies in the country from which an exotic pest has been introduced. Following the discovery of a potential bio control agent, it undergoes extensive evaluation to insure that its ecology and host range are compatible with the community to which it will be introduced and that it will not become a pest once it is released. Suitable candidates are reared and released in the new habitat with an intention that they will become established and suppress the pest population.

2. Augmentation

This may be defined as the effort to increase population of natural enemies either by propagation and periodic release or by environmental manipulation. Natural enemies that are unable to survive and/or persist in a new environment can sometimes be reared in large numbers and periodically released to suppress a pest population.

Two approaches in augmentation

- **Innoculative release:** Innoculative releases are used in case where control is expected from the progeny and subsequent generations and not from release itself. In some cases, small numbers of a beneficial species are released in several critical locations to suppress local pest outbreaks. These may be made as infrequent as once in a year to reestablish.

- **Inundative release:** In other cases, larger numbers are released in a single location to flood the pest population with natural enemies. It involves mass culture and release of natural enemies to suppress the pest population. Here the control is from release itself.

3. Conservation

It may be defined as action to preserve and increase natural enemies by environmental manipulation. A variety of management activities can be employed to optimize the survival and / or effectiveness of natural enemies.

- Reducing insecticide applications to avoid killing of natural enemies,
- Providing shelter,
- Providing over-wintering sites, or alternative food sources to improve survival of beneficial species.
- Avoidance of cultural practices which are harmful to natural enemies.
- Use of cultural practices which favors survival and multiplication of natural enemies.
- Provision of food like pollen and nectar for adult stages of natural enemies.
- Suppression of ant to increase the efficiency of predators.
- Go for trap cropping for increasing the efficiency of predators and parasitoids.

Physical Control

The physical method includes temperature, sound, controlled atmospheres and radiations which are utilized for pest control by affecting them physically or altering their physical environment.

Temperature

- Stored grain insects will die at temperature less than 13°C or higher than 35°C.
- For majority of stored grain pests, an exposure of 14°C will cause complete elimination of the pest.

Sound

- Insects produce sound which may serve as a warning device to communicate, find mates and isolate species.

- Insects produce sound by using a variety of mechanisms, including beating wings during flight.
- Low frequency sound waves cause adverse effect on development of insects. Sound produced by male and response of female of a species to the sound can be utilized for their control.

Controlled Atmospheres

- Use of controlled atmospheres to manage insect pests has gained momentum in recent years.
- Carbon dioxide is toxic to insect, but its action is low. Eggs and adults of pulse beetle die when exposed to 100% Carbon dioxide at 32°C and relative humidly of 70%.
- Carbon dioxide under high pressure is found to be effective against stored grain pests.
- Carbon dioxide and nitrogen treatment have been found effective for grain beetle. A nitrogen rich atmosphere effectively controls all stages of fruit fly.

Irradiation

- Ionizing radiation consists of gamma rays and electron beam irradiation.
- Non-ionizing-radiation includes radio waves, infrared waves, visible light and microwaves.

Mechanical control

- The reduction or suppression of insect populations by means of manual devices is referred to as mechanical control.
- The mechanical control involves the use of the following tactics:

Hand Picking

- Collection and destruction of egg masses and destruction of infested cane stocks harboring larvae of borers is useful for reducing the incidence of insect pests.
- Hand picking is also generally useful for the management of hairy caterpillars, leaf rollers, leaf webbers, tobacco caterpillar, cabbage butterfly, mustard sawfly, Epilachina beetle, white grubs etc.

Mechanical exclusion

- Mechanical exclusion consists of the use of devices by which barriers are used which can stop the pest from being able to reach insects and physically prevented from reaching crops and agricultural produce.

- The application of a fluffy cotton band 6" wide, or a band of a sticky material or a band of slippery sheets like alkathene around the tree trunk of a mango tree to prevent the upward movement of the mango mealy bug.

- Screening windows, doors and ventilators of the house to keep away houseflies and mosquitoes, bugs etc.

- Wrapping individual fruit of pomegranate and citrus with butter paper envelope to save them from attack of the Annar butterfly and fruit sucking moths respectively.

Digging trenches

Digging of 30 to 60 cm deep trenches or erecting 30 cm high tin sheet barriers around field is useful for protecting them from moving swarms of locusts and hairy caterpillars.

Lighting

- Using red light in the monsoons to keep away most of insects, and to keep the field well lit with white light at night to protect it against certain insects.

- Light reflection by aluminum foil is effective against aphids.

Traps

- Used for monitoring initial infestation and the periodicity of pest activities. Various types of traps have been devised for collecting and killing different types of insects:

    i)   The cricket trap: A deep cylindrical vessel containing beer as bait and having wooden splinters to aid crickets to reach the bottom.

    ii) House-fly trap: A box, containing a piece of stale cake, with a side opening for the insects to get in only to be trapped in a wire gauze cage on the top

    iii) Light traps: Light traps for attracting and mass killing of several species of moths and beetles were used as a control measure before the advent of synthetic organic insecticides. The traps could still be useful for monitoring the population of important pests in an area. Trapping of adults through light traps has proved encouraging in controlling top borer, root borer and white grub damage in sugarcane, red hairy caterpillar and beer beetle.

    iv) Fermentation traps: Pheromone baits used in traps are being used for reducing lepidopteron (moths and butterfly) pests. Moths are naturally attracted to molasses, fermenting fruit, tree sap, honeydew and flower nectar.

v) Coloured traps: Different insects respond to different colours. The selection of the colours depend upon the position of traps, physiological stage of insect and quality of the incident wavelengths hitting the traps.

Banding of fruit trees

* This is effective against caterpillars, ants and mealy bugs which crawl up the trunk from the soil to the shoots.

* The banding material like grease smeared around trunk trees arrests the crawling insects below the banded part where either they die of starvation or by getting entangled with the sticky banding material.

Clipping, Pruning and Crushing

* Pruning and destruction of infested shoots and floral parts is effective in checking the multiplication of scales, mealy bugs and gall midges attacking fruit trees like grapes, citrus, ber, fig, custard apple etc.

* Pruning of infested material is useful approaches for the management of mustard aphid infesting mustard crop involving clipping and destruction of aphid-infested twigs.

Beating and Hooking

* Killing houseflies with fly wrappers and locusts with brooms or thorny bushes is effective. On coconut palms, the borer can be picked out of the holes with the help of crooked hooks made of iron.

Shaking or Jarring

* Shaking small trees and shrubs, particularly early in the morning in the cold season when the insects are benumbed, and collecting them in open tubs containing kerosinized water or simply burying them in pits iseffective against locusts and the defoliating beetles.

Sieving and Winnowing

* These are commonly employed against insect pests of stored grains. A good number are removed with these operations, particularly the grubs of *Tribolium* and *Trogoderma*, which infest wheat.

Bio-pesticidal control strategies

Biopesticides includes:
* Botanical pesticides
* Microbial
* Entomopathogenic nematodes

Botanical pesticides

- Botanical insecticides are naturally occurring chemicals (insect toxins) extracted or derived from plants that have naturally occurring defensive properties. They are also called natural insecticides.

Table 1: List of Botanical insecticides, sources and mode of action.

Botanical insecticides	Sources	Mode of action
Pyrethrins	*Chrysanthemum cinerariaefolium*	Disrupting the sodium and potassium ion exchange process
Neem	*Azadirachta indica*	Feeding deterrent
Rotenone	*Derris* species	Inhibitor of cellular respiration
Sabadilla	*Schoenocaulon officinale*	Affect neurotransmitter action
Ryania	*Ryania speciosa*	Slow-acting stomach poison
Nicotine	*Nictiana tabacum*	Affects bonding to acetylcholine receptors
d-Limonene and Linalool	Citrus fruit peels.	Paralysis

Microbial pesticides

i) Bacteria

- Some entomopathogenic bacteria present in soil are *Bacillus* and *Paenibacillus* are pathogenic to coleopteran, dipteran and lepidopteran insects.

- There are spore-forming bacterial entomopathogens such as *Bacillus* spp., *Paenibacillus* spp. and *Clostridium* spp. and non-spore forming ones that belong to genera *Pseudomonas* spp., *Serratia* spp., *Yersinia* spp., *Photorhabdus*spp. and *Xenorhabdus* spp.

- The best known and most widely used is *Bacillus thuringiensis* subspp. For example *Bacillus thuringiensis* subsp. *aizawai* and *Bacillus thuringiensis* subsp. *kurstaki* are effective against caterpillars.

ii) Fungus

- Fungal pathogens have a broad host range and are especially suitable for controlling pests that have piercing and sucking mouth parts because spores do not have to be ingested. For example the target pests of the entomo-pathogenic fungus *Metarhizium anisopliae* is Coleoptera, Lepidoptera, Hemiptera and Orthoptera.

iii) Viruses

- Entomopathogenic viruses need to be ingested by the insect host and therefore are ideal for controlling pests that have chewing mouthparts.

- Several lepidopteran pests are important hosts of baculoviruses including nucleo polyhedron viruses (NPV) and granulo viruses (GV).

- Virus particles invade the nucleus of the midgut, fat body or other tissue cells and cause damage to the pest. For example *Helicoverpa armigera* nucleo polyhedron virus (HaNPV) act against the *Helicoverpa armigera*.

Entomopathogenic nematodes

- Entomopathogenic nematodes are microscopic, several species of *Heterorhabditis* and *Steinernema* are available in multiple commercial formulations.

- Infective juveniles of entomopathogenic nematodes actively reach to their hosts and enter through natural openings such as mouth, spiracles and anus or inter segmental membrane. Once enters into host body, the nematodes release symbiotic bacteria that multiply rapidly produce toxins that kill the host within 24-48 hours through bacterial septicemia. For example *Heterorhabditis* spp. carries *Photorhabdus* spp. bacteria.

Table 2: Biopesticides based on various entomopathogenic microorganisms and their target pests.

Entomopathogens	Trade name of biopesticides	Target pests
Bacteria		
Bacillus thuringiensis subsp. *aizawai*	Agree WG	Lepidoptera
Bacillus thuringiensis subsp. *kurstaki*	Dipel ES, Costar and Thuricide	Lepidoptera
Bacillus thuringiensis subsp. *israelensis*	Mosquito beater WSP	Diptera
Bacillus thuringiensis subsp. *tenebrionis*	Novodor FC	Coleoptera
Fungi		
Beauveria bassiana	BotaniGard ES and Myco-Jaal	Pest of Acarina, Coleoptera, Diptera, Hemiptera, Lepidoptera and Thysanoptera etc.
Hirsutella thompsonii	ABTEC Hirsutella	
Isaria fumosorosea	NoFly WP	
Lecanicillium lecanii	PhuleBugicide	
Lecanicillium longisporum	Vertalec	
Viruses		
Granulovirus (*Cydiapomonella* GV)	CYD-X and MADEX HP	Lepidopterons
*Helicoverpa zea*NPV	Gemstar LC	
Spodopter aexigua NPV	Spod-X LC	
Nematodes		
Heterorhabditis bacteriophora	Nemasys and Terranem	Several orders of soil borne pests
Heterorhabditis heliothidis	Double-Death	
Steinernema carpocapsae	Ecomask and NemAttack	
Steinernema feltiae	Econem and Scanmask	

References

Attra //Biointensive Integrated Pest Management (www.attra.ncat.org)

Baer, C. F., D. W. Tripp, A. Bjorksten and M. F. Antolin. (2004). Phylogeography of a parasitoid wasp (*Diaretiella rapae*): no evidence of host-associated lineages. *Molecular Ecology*, 13: 1859-1869.

Benbrook, C.M. (1996). Pest Management at the Crossroads. Yonkers, N.Y: Consumers Union.

Glare, T.R., Caradus, J., Gelernter, W., Jackson, T., Keyhani, N., Kohl, J., Marrone, P., Morin, L., Stewart, A., (2012). Have biopesticides come of age? *Trends Biotechnol*. 30: 250-258.

Hilker M., McNeil J. (2008). Chemical and behavioral ecology in insect parasitoids: how to behave optimally in a complex odorous environment. In: Wajnberg E, Bernstein C, van Alphen J, (Editors). Behavioral Ecology of Insect Parasitoids. Blackwell Publishing, Oxford, UK. 92-112.

Krishna Moorthy and Krishna Kumar, (2004). Integrated Pest Management in Vegetable Crops Eds. Birthal, S.P and Sharma,O.P. *Proceedings 11 Integrated Pest Management in Indian Agriculture*. 95-108.

Singh, H.S., V. Pandey, G. Naik and Vishal Nath, (2007). Developing trellis system for efficient pest control in bitter gourd under coastal ago-ecosystem of Orissa. *Vegetable Science*. 34(2): 140-144.

Srinivasan, K. and P.N. Krishna Moorthy. (1993). Evaluation of neem products and other standard chemicals for the management of major pest complex on cabbage: Comparison between standard spray regime and IPM involving mustard as a trap crop. In: Neem and Environment Vol. 1 (eds. R.P. Singh, M.S. Chari, A.K. Raheja and W. Kraus). New Delhi: Oxford & IBH Publishing Co.

Fruit Crops

3

Biointensive Integrated Pest Management of Litchi

Jaipal Singh Choudhary, Kuldeep Srivastav and Bikash Das

Introduction

Litchi (*Litchi chinensis* Sonn), like most fruit tree crops, is usually attacked by two or three key pests, several secondary pests and a large number of occasional pests in localized areas where it is grown. Litchi is an important subtropical evergreen fruit crop belongs to family Sapindaceae. It is known as queen of the fruit due to its attractive deep pink/red colours and flavoured juicy aril (Singh *et al.*, 2012).

Insect pest attacks on litchi, in China there are 193 species of litchi pests in two classes, 11 orders, and 57 families. In India, nearly 42 insects and mite species reported to attacking on trees and fruits of litchi at different stages of growth (Singh *et al.*, 2012). The majority belong to the Lepidoptera and Coleoptera, followed by the Homoptera and Hemiptera. However, only about three or four key pests including fruit and flower borers, sucking bugs, stem borers and erinose mites which require regular control, and the species involved vary between orchards. Secondary pests viz., fruit-piercing moths, defoliators etc generally occur at sub-economic levels, but can become serious pests as a result of changes in cultural practices and litchi cultivars or because of indiscriminate use of insecticides against a key pest. Recently, litchi fruit borer and litchi leaf roller have acquired the status of major pests and now, litchi looper, litchi bug and bag worm are emerging pests of litchi (Choudhary *et al.*, 2013; Singh *et al.*, 2014). Occasional or incidental pests also can cause economic damage only in localized areas at certain times; the great majority of pests reported here fall within this last category.

Earlier controlling of litchi pests simply involves the pre-determined spraying if insecticides without the monitoring of pest population dynamics. Farmers prefer the chemical insecticides spray due to their quick and visible impacts. Chemical insecticides spraying are also justifiable because of high price of litchi in the market. Generally 5-6 insecticides spray in a year are common to get a good crop of litchi (Srivastava *et al.*, 2016). Use of insecticide results in pollution of the surrounding environment and leaves orchards in poor condition, with degraded biodiversity and fewer natural enemies, and makes them more susceptible to outbreaks of minor pests which may reduce or completely destroy the crop in some years (Li *et al.*, 2014). Keeping the above facts in mind, we in this chapter focus on Biointensive insect pest management in the litchi orchards to get pesticide residue free litchi fruits.

1. Litchi Fruit Borers

Conopomorpha sinensis Bradley & *C. litchiella* Bradley (Lepidoptera: Gracillariidae)

Symptoms of damage: The newly hatched larvae start feeding in to the seed/ nut by boring through pulp. Black spot near the pedicel is the identification mark of fruit infestation. Larvae usually feed on the developing seed of fruits. Infested fruits give the appearance of excreta/larva, when fruit is cut/ opened after ripening. Ultimately fruit drop down before maturity as result of borer infestation. In this way, larvae of these borer causes direct damage to litchi fruits (Fig. 1). During July-October month, indirect damage causes by larvae through making mines in to young leaves and shoots resulting in failure of shoots to bloom. As a result of mines branches withers and drop. Almost all the litchi varieties are susceptible to borer infestation. Fruit loss varies from 30-90 per cent due to borer infestation.

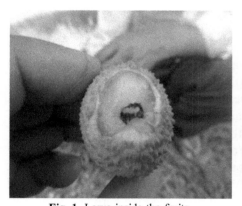 

Fig. 1. Larva inside the fruits   **Fig. 2.** Pupae on the leaf tip

Seasonal incidence: Intermittent rains and blowing of easterly winds during seed development are the congenial conditions for litchi fruit borer population build up. Fruit infestation rate and fruit growth also showed sigmoidal and a highly significant positive correlation (Schulte *et al.*, 2007). Maximum fruit damage is recorded April-May in the fruits and highest population of pest in August-

September months usually recorded on younger shoots. It is also reported that from October to March, pest population was not found in orchards but reappeared in the month of April (Hameed *et al.*, 1992).

Biology: After mating, adults lay their scale like minute eggs singly and total up to 50 on the under surface of the leaf or near the calyx on the pericarp of litchi fruits. Eggs are yellow-orange in colour and flattened in shape. Incubation period lasts for 4 to 5 days. There are five larval instars. First instar

Fig. 3. Adult *C. sinensis*

larvae are milky white with distinct light brown head. Larvae are yellowish in color with brownish head. Fully grown larva is yellowish in color. Larval period is 7-8 days. Fully grown larva is yellowish in color and measured 8.0±0.1 mm long with 0.8±0.01 mm wide. Fully grown larvae come out of the fruit and pupate on the leaf surface under spinning cocoon (Fig. 2). The color of pupa is greenish on dorsal side and greenish yellow in ventral side. Pupal period is 6-7 days. Adults of *C. litchiella* species are a small brown moth characterized by bright yellow patches at the tips of the forewings while adults of *C. sinensis* are greyish brown in color with wings having zigzag colored pattern (Fig. 3). Longevity of the adults varies from 3.12 to 6.84 days.

Biointensive integrated management

- Pheromone based trapping should be placed in the litchi orchards for pests monitoring.

- If adult traps, release of *Trichogramma chilonis* @50000 eggs/ha immediately in the orchards is advised.

- Paper bagging of fruit bunch after fruit setting may be keeps out the pest incidence on fruits. Bagging also improved fruit colour and quality.

- Trees should be pruned after harvest to remove harbour pupae of the pest.

- Organically fruit and shoot borer can be managed by using neem oil as a deterrent, hence spraying of neem oil (4 ml/l) before flowering may be done to avoid egg laying by fruit borer.

- Two sprays with panchgavya (30 ml/l) made up of cow ghee, urine, dung, curd milk along with banana and sugarcane juice at clove stage and colour break stage or two spray with biodynamic pesticide (50ml/l) made up with cow urine, cow-dung, chopped leaves of neem/Calotropis (madar) decomposed in water at aril (pulp) development stage and about 10 days before expected fruit harvest can also be good alternate of chemical insecticides.

• Adoption of good orchard management practices like pruning of infested twigs in June, manuring of plants with 4 kg castor-cake and 1 kg neem-cake in the 2nd week of July and prophylactic spray of neem based formulations (4 ml/L) during new flush emergence (September-October) improve the effectiveness of organic schedule (Srivastava *et al.*, 2015).

• If needed, spray the orchard with safe insecticide, spinosad 1ml/4 l of water after fruit set and second spray may be given before 15 days harvest.

2. Litchi stink bugs

Tessaratoma javanica (Thunberg), *Tessaratoma papillosa* (Dury), and *T. quadrata* Distant *(Hemiptera: Tessaratomidae)*

Stink bugs are regular pests in all the litchi growing areas in India causing huge economic losses in quality and quantity of the litchi (Choudhary *et al.*, 2013). *T. javanica*, is a major pest species of stink bug in India. Both adults and nymph suck cell sap mostly on tender plant parts such as growing buds, leaf petioles, fruit stalks and tender branches of litchi tree (Fig. 4).

Fig. 4. *Tessaratoma javanica*

Symptoms of damage: The adult as well as nymphs both causes direct damage by sucking the sap on the flowers and fruits, which may wilt and die and ultimately to fall down. When disturbed the stink bug secrete an acid fluid which induces wilting of young leaves and causes brown spots on old leaves and fruits reducing their economical value. In sever attack it leads to heavy fruit dropping and ultimately total damage to the litchi crop. The bugs when crowd on the developing fruit, it causes the fruits to fall a couple of week later. The losses go up to 70-90 percent (Choudhary *et al.*, 2013).

Seasonal incidence: The bug infestation is observed throughout the year but it is high in summer, moderate during the post-rainy season and low in the rainy season. The peak activity of the bug is from the last week of April to the last week of August after that it disappears from the orchard and undergoes hibernation in adult stages, in the northwestern part of India. In eastern India peak incidence of this bug has been observed from February to September (Choudhary *et al.*, 2015).

Biology: The population of litchi stink bug starts rising with the emergence of inflorescences and reaches on peak at full bloom stage of flowers which coincide with pollinators activity (Choudhary *et al.*, 2013). Adult bugs of *T. javanica* lays globular and off pink eggs, but it takes longer (180 days) during hibernation period (August to February). Oviposition occurs 13 days after mating, eggs are usually laid in clusters of 14 eggs, arranged in 3–4 rows. Eggs are preferably laid under leaf surface, but also observed on inflorescences, fruits and over the already laid eggs. The mean fecundity per female is observed 13.4 egg clusters. Immature stages are gregarious. The bug passes through five instars, which are sub rectangular and dark brick red except first instar, which are nearly sub-rectangular. Newly emerged nymph is dirty white and soft bodied insect but colour changes to yellow-red after few days. The life span of all stages completes in 13, 60, 60 and 75 days of egg, nymph, male and female adults, respectively (Parveen *et al.*, 2015).

Biointensive integrated management

- Biointensive management strategy for *T. javanica* involves reducing the overwintering adult populations as much as possible before egg laying commences by manual removal of adults.

- The egg masses of litchi stink bug are parasitizing by egg parasitoids, viz. *Anastatus colemani* (Hymenoptera: Eupelmidae) and *Ooencyrtus phongi*. These two parasitoids offer great potential for the biological control of litchi bug and parasitizing to eggs ranged from 12.9 to 52.9% and 20.6 to 55.7% *A. colemani* and *O. phongi*, respectively.

- Three egg parasitoids, two are from the family Eupelmidae and one is from the family Encyrtidae of order Hymenoptera namely *Anastatus bangalorensis* Mani & Kurian, *Anastatus acherontiae*. The maximum activity of parasitization was recorded in month of March laid eggs. The cumulative parasitization of eggs may reach up to 70 to 90 per cent of eggs laid late in the season in favorable weather conditions. These reported egg parasitoids of litchi stink bug could be beneficial in the integrated management of the pest if mass reared and released in the litchi orchards as usual practice in China. Mechanical exclusion of parasitized eggs collected from natural incidence condition may also improve the parasitization.

- Use of judicious and timely application of insecticides may protect the pollinators and egg parasitoids.

- Spray of lambda cyhalothrin+diclorvos (DDVP) @ 0.5+1.0 ml/litre of water in the last week of December month, spray of dimethoate @0.05% in second week of February on immigrated population after panicle emergence and before flower opening, spray of imidacloprid @0.005 against first instar nymphs at100% fruit setting stage and fourth spray of acephate @0.1% at pea stage of fruits was found effective and economical viable. Spraying is conducted in early to mid-March and again in early May, before flowering and during the early fruit stage, respectively. This protects both honeybees, during pollination of litchi flowers, and egg parasitoids, which is mass released during the year.

3. Litchi erineum mite

Aceria litchii (Keifer) *(Acari: Eriophyidae)*

Mite is the major pests in the litchi growing parts of the India. The other species which attack on litchi reported as *Oligonychus mangiferus (Acari: Tetranychidae)* is a minor pest.

Symptoms of damage: Litchi mite is a destructive and endemic pest causing curling of leaves. The curling of leaves and chocolate-brown growth on the ventral surface of leaves at later stage is visible (Fig. 5).

Fig. 5. Mite infestation (Velvety growth)

Seasonal incidence: The incidence of litchi mite can be observed round the year on litchi trees. The highest population (18.1 per cm^2 leaf surface) of mite is observed during March, while lowest population (1.4 per cm^2 leaf surface) is observed during November.

Biology: The adults start multiplying from the end of February or March and the peak activity is noticed around July. Both nymphs and adults are vermiform. The female lays eggs, singly at the base of the hair on the lower surface of leaves. The eggs are small, round and whitish hatching within 2-3 days and newly emerged nymphs feed on soft leaves. The nymphs and adults are similar in appearance, whitish and four legged, however, nymphs being smaller and have less number of lateral setae. Life cycle is completed within 8-20 days with 10-12 generations in a year (Prasad and Singh, 1981).

Biointensive integrated management

- The leaves should be checked regularly for symptoms over summer and autumn.

- Branches infested with the mite should be cut off.

- After harvesting in June, infested branches must be removed.

- Manuring the litchi trees with 4 kg of castor and 1 kg of neem cake, when applied at root zone after the first shower monsoon have been proved effective in reducing mites infestation. Layer saplings may be sprayed with Dimethoate @ 2ml/l before leaving the nursery.

- Prior to planting out, the operation should be repeated twice at 10-14 day intervals.

- In September-October, trees must be treated just prior to vegetative flushing with Dimethoate @ 2 ml/l, spraying should be repeated after two weeks. In the month of February spraying with propargite @ 2.5 ml/l before flowering have been found useful.

- Need based spraying of buprofezin @1.5 ml/l may done just after fruit setting to manage mites. Diafenthiuron, Hexythiazox, Fenazaquin, Spiromesifen, Milbemectin, Dicofol, Azadirachtin are also reported to effective against mites pests.

4. Bark Eating Caterpillars

Indarbela tetraonis Moore and *Indarbela quadrinotata* Walker *(Lepidoptera: Cossidae)*

Symptoms of damage: Pest damage can be characterized by the presence of long-winding, thick, blackish or brownish ribbon-like masses composed of small chips of wood and excreta. Damage is caused by caterpillar, which bore into trunk, main stems and thick branches of litchi tree and destroying xylem tissues resulting into poor growth and fruiting of the tree. Caterpillars remain within the tunnel inside the stem during day, come out in night and feed upon the bark.

There is only one caterpillar in each hole, and there may be 2-16 holes in each tree, depending upon the intensity of infestation and age of tree. Older and uncared trees are more affected by the pest. Attack of pest weakens the stem, resulting in drying of the branches and finally tree may die itself.

Fig. 6. Larva and larval frass of bark eating caterpillar

Seasonal incidence and biology: Female moth lays 250-300 eggs in loose bark, cuts and crevices of bark in clusters in early June. Eggs hatch in 15-25 days. Larvae that hatch out initially feed on the bark and subsequently bore into the trunk. The fully grown caterpillars are with dark brown head and dirty brown body. The tunnel entry remains closed with a frass covering which is drawn out into a sleeve through which the larva moves (Fig. 6). The full grown caterpillars develop after 4-5 moultings with 9- 10 months. Pupation occurs within the tunnel, with the cephalic end of the pupa slightly protruding outside. The pupal period lasts for about 15-25 days. Adult moths start emerging during May to July month. Adult moths are stout, light grey in colour with dark brown dots (Hameed *et al.*, 2001).

Biointensive integrated management

- Incidence of pest can be minimize through removal and destroy of dead and severely affected branches of the litchi tree.

- The caterpillars can be killed by inserting an iron spoke into the tunnels.

- This insect has also been successfully controlled by injecting kerosene oil into the tunnel by means of a syringe and then sealing the opening of the tunnel with mud.

5. Leaf roller

Dudua aprobola (Meyrick) *(Lepidoptera: Tortricidae)*

Symptoms of damage: This pest is a very serious litchi pest in Bihar and surrounding areas. The symptoms of leaf damage by the larvae are manifested through rolling of tender leaves and feeding inside (Fig. 7). As a result of larval damage, the infested twigs distort and wither. Leaf damage goes up to 70 per cent. It may attack flowers also.

Seasonal incidence: The incidence of leaf roller can be observed on litchi trees throughout the year. The breeding season of the leaf roller on litchi plants is from August to February month when new leaf flush is available. Maximum population is observed from July-February (Ray and Mukherjee, 2012).

Biology: After the emergence of adult moths, mating and oviposition takes place during night time. The female moth can lays creamy white 150-200 eggs under the surface of newly emerged tender leaves. After hatching within 2 to 8 days, larvae undergo five moults before pupation. Colour of larvae head is red-brown, brownish or yellowish and flattened in shape (Fig. 8). The larval and pupal period varies from 7 to 14 days and 7 to 24 days, respectively. Adult moths have forewing colour ranging from an overall brownish shade to dark greyish-blue. The insect completes its life cycle within 21.6-31.0 days in July-August and 46.0-46.5 days in February-March.

 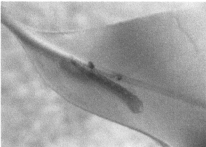

Fig. 7. Rolled leaf   **Fig. 8.** Larva

Biointensive integrated management

- Removal of affected flush and summer ploughing can reduced the pest population below threshold level. The rolled leaves that contain larvae may be removed manually during light infestations to check further infestation.

- If necessary, spray of nimbecidine 300 may be applied when 20 per cent of leaf flushes are infested to minimize damage to young trees or at critical periods of leaf growth in older trees.

6. Litchi Looper

Perixera illepidaria Guenée (*Lepidoptera: Geometridae)*

Symptoms of damage: It attacks tender leaves in mass and defoliate the new shoots. Litchi trees have been attacked by this pest results very poor flowering in bearing orchards and retard the growth in young saplings in subsequent season. When infestation is high larvae feed voraciously on the lamina of young leaves leaving only the midribs and veins (Fig. 9). Also the larvae sometimes feed on tender shoots.

Fig. 9. Larva

Seasonal incidence: Incidence of this pest is normally observed from July December however, highest population is found in month September- October.

Biology: Female lays eggs on the under surface of the leaves after the mating. Egg hatches within 4-5 days. Larvae of litchi looper show variation in the color from black to dark brown. Larval and pupal periods are completed in 8-9 and 5-7 days, respectively. The adult male is pinkish brown, while females are rather uniformly pinkish. The wings of the adult have 2 rows of dark brown spots on the dorsal surface, the first row being just near the distal edge of each wing. The margins of both fore wings and hind wings are serrated having tufts of fine hairs. The developmental period from larva to adult is completes within 15-19 days. Five to six overlapping generations are completed in a year (Kumar *et al.*, 2014).

Biointensive integrated management

- Larvae can be manually picked by hand to reduce the incidence at early stage of infestation. Loopers pupate at the leaf surface and can be seen easily therefore, may also be removed manually through pruning.
- Spraying of safe insecticides i.e., Indoxacarb 15.8 EC at 0.7 ml/lit or azadirachtin 10,000 ppm at 2 ml/lit water is effective against the pest when it is requires.

7. Green looper

Thalassodes pilaria Guenee *(Lepidoptera: Geometridae)*

Symptoms of damage: The larvae of this pest are feeding on tender foliage and resemble a green stick similar to the midrib of leaves or thin shoots that served as a camouflage for the pest.

Seasonal incidence: The infestation of this pest is seen from July to December on new vegetative shoots.

 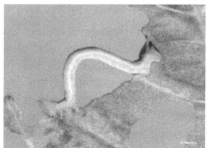

Fig. 10. Adult   **Fig. 11.** Larva

Biology: Larval period completes within 9-13 days, and the pupal period from 6 to 7 days. The adult moths with fully spread wings are 27-30 mm wide. The adult moths have sea-green colour wings which are semitransparent having fine white lines (Fig. 10 & 11). Fore wing is broad, triangular and leading edge was yellowish brown, crossed by two distinct white lines and a number of faint transverse marks which is also white. The hind wings have an angular margin. Underside of the moth is greenish white (Kumar *et al.*, 2013).

Biointensive integrated management

Management for green looper is same as litchi loopers.

8. Shoot borer

Chlumetia transversa (Walker) *(Lepidoptera: Noctuidae)*

Symptoms of damage: It is a minor pest on litchi but in case of severe infestation the sap movement is interrupted and the tree ceases to flush. Larvae of this moth bore into the young shoot resulting in dropping of leaves and wilting of shoots. Female moths lay egg on tender leaves. After hatching, young larvae enter the midrib of leaves and then enter into young shoots  through the growing points by tunneling downwards (Fig. 12). Larvae also bore into the inflorescence stalk.

Seasonal incidence: This pest is observed to damage new flushes of litchi throughout the country usually active from August to October (Rai *et al.*, 2002).

Biology: The Caterpillars of this species are pale brown with a black head. The adult moth is brown with shaded bands across the forewings, and a pale patch near the tip of each forewing. The hind wings are plain brown. There are four overlapping of the pest in a year and it overwinter in pupal stage.

Biointensive integrated management

- The attacked shoots may be clipped off and destroyed to reduce further incidence.

- Larval parasitoids, *Megaselia chlumetiae*, *B. greeni* etc. may act as good biological control agent for *C. transversa*.

- Two spraying of cypermethrin 25EC (0.4 ml/L) at 7 days interval can effectively control the pest. A total of 2 to 3 sprays may be given depending on the intensity of infestation.

9. Leaf cutting weevils

Reported species of leaf cutting weevils on litchi are *Myllocerus undatus* Marshall, *Myllocerus discolor* Boheman, *M. delicatulus* Boheman, *M. undecimpustulatus* Faust, *M. discolor* Boheman, *M. dorsatus* (Fabricius)- *Amblyrrhinus poricollis* Schönherr *(Coleoptera: Curculionidae)* and *Apoderus blandus (Coleoptera: Attelabidae)*.

Symptoms of damage and identification

Adult weevils congregate on the tender leaves and nibble irregular holes on the leaves and sometimes consume the entire leaf leaving the midrib only (Fig. 13). Adult of grey weevils has long snout with grey colour, though poor flier but very active feeder on the leaves of litchi. Adult of *A. blandus* are bright brownish red in colour. The ventral side (abdomen) is looking pale brown, and parts of the mouth and claws are brownish red. The damage of red weevil is

Fig. 13. Damage by shoot borer on new flush

more severe at the time of shoot emergence as it prefers newly leaves therefore; newly established orchard/nurseries are more vulnerable for pest attack (Mazumder *et al.*, 2014).

Biointensive integrated management

- The grubs of these weevils feed on organic matter in the soil below the canopy, hence, ploughing and exposing these grubs reduces the problem.
- Hand picking of the adult weevils reduces their incidence to some extent
- Adult weevils can be removed from plants by vigorously shaking a branch over an open, inverted umbrella. The collected weevils can then be dumped into a container of soapy water.
- Spraying of Buprofezin 25 SC @ 1ml/l is highly effective against weevils.

10. Lac insect

Kerria lacca Kerr. & *K. albizziae* (Green) *(Homoptera: Lacciferidae)*

Symptoms of damage: It is known to attack litchi plants in different part of world (Sharma *et al.*, 2006). Nymphs feeding by piercing the shoot with their proboscis. Once the proboscis is thrust into a shoot, the nymphs get settled down, they secrete resin over their body. After successfully colonized by the insect, intermediate secondary growth, viz., non-formation of a thick layer of cork tissues can be observe (Fig. 14). Lac insect is found damaging litchi not only covering the stem with its resinous secretion but also twigs and leaves with honeydew on which sooty mold fungus developed and interfered with photosynthesis and growth of plant.

Seasonal incidence and biology

The female insect lays eggs inside the encrustation. They hatch almost immediately, and

Fig. 14. Lac insect infestation on litchi

the nymphs crawl out of the cell. Nymphs walk on the shoots, move mostly upward towards the tender branches and settle on them. They start feeding by piercing the shoot with their proboscis. Once the proboscis is thrust into a shoot, the nymphs get settled down, they secrete resin over their body. After successfully colonized by the insect, intermediate secondary growth, viz., non-formation of a thick layer of cork tissues can be observe. The resin glands are distributed all over the insect.

Biointensive integrated management

• The lac insect is known to be attacked by parasitoids viz., *Aprostocetus* (Syn. *Tetrastichus*) *purpureus* (Cam.) (Eulophidae), *Coccophagus tschirchii* Mahd. (Aphelinidae), *Tachardiaephagus tachardiae* How (Encyrtidae) and predators viz., *Eublemma amabilis* Moore (Noctuidae) and *Pseudohypatopa* (=*Holcocera*) *pulverea* Meyr. (Blastobasidae) in nature which keep them under control (Sharma *et al.*, 2006).

11. Thrips

Dolicothrips indicus Hood and *Megalurothrips usitatus* Begnall & *M. distalis* Karny *(Thysanotera: Thripidae)*

Dolicothrips indicus and *M. usitatus* (Nayar *et al.*, 1992) cause damage to flowers, while *M. distalis* Karny attacks on the leaves (Singh *et al.*, 2012). Adult thrips are small, elongated and fast moving, approximately 2 mm in length with four narrow fringed wings. The life cycle is completed in about 15-18 days. Nymphs are similar to adults but are without wings.

Fig. 15: Thrips damaged flowers of litchi

Adult thrips live for about 10-12 days. Damage is caused both by nymphs and adults by rasping the lower surface of the leaf with their stylets and sucking the oozing cell sap from wound. Species which attack on leaves are usually found on the leaf tips. Attack by thrips on litchi causes drying of leaves and flowers ultimately fall down (Fig. 15). Thrips have not been reported to cause heavy damage to litchi crop in India.

Biointensive integrated management

• Sticky traps and tapping can be used to determine the presence or absence of thrips in litchi rather than for the control measures.

• If incidence is there, use of *Verticillium lecanii* (Zimm.) or *Beauveria bassiana* (Balsamo) @ 5 mL or 5g/L or Thiacloprid 21.7 SC or Fipronil 5% SC or Spinosad 45% SC is effective against thrips in litchi.

References

Choudhary, J.S., Prabhakar S.C., Moanaro and Maurya, S. (2015). New Record of the Litchi Stink Bug, *Tessaratoma javanica* (Thunberg) Egg Parasitoids and their Natural Control Effect in Litchi Orchards from India. *Entomologia Generalis*, 35(3): 187-197.

Choudhary, J.S., Prabhakar S.C., Moanaro, D.B., and Kumar, S. (2013). Litchi stink bug (*Tessaratoma javanica*) outbreak in Jharkhand, India, on litchi. *Phytoparasitica* 41: 73-77.

Hameed, S.F., Sharma, D.D. and Agarwal, M.L. (1992). Integrated Pest management in litchi. Proceedings of National Seminar on Recent Developments in Litchi Production held at R.A.U., Pusa. pp. 38.

Hameed, S.F., Singh, P.P. and Singh, S.P. (2001). pests. In: Chauhan, K.S. (ed). Litchi: Botany, Production, Utilization. Kalyani Publishers, Ludhiana, 228 pp.

Kumar, V., Venkatarami Reddy, P. and Anal, A.K.D. (2013). Report on the occurrence and biology of *Thalassodes pilaria* Guenee (Lepidoptera: Geometridae) on litchi (*Litchi chinensis* Sonn.) in Bihar, India. *Pest Management in Horticultural Ecosystems* 19(2): 245-247.

Kumar,V., Venkatarami Reddy, P., Anal, A.K.D. and Nath, V. (2014). Outbreak of the looper, *Perixera illepidaria* (Lepidoptera: Geometridae) on litchi, *Litchi chinensis* (Sapindales: Sapindaceae) - a new pest record from India. *Florida Entomologist* 97(1): 22-29.

Li, S.D., Liao, C., Zhang, B.X. and Song, Z.W. (2014). Biological control of insect pests in litchi orchards in China. *Biological Control,* 68: 23-36.

Mazumder, N., Dutta, S.K., Bora, P., Gogoi, S. and Das, P. (2014). Record of litchi weevil, *Myllocerus discolor* (Coleoptera: Curculionidae) on litchi (*Litchi sinensis* Sonn. (Sapindaceae) from Assam.*Insect Environment* 20(1): 29-31.

Nayar, K.K., Ananthakrishanan, T.N. and David, B.V. (1992). General and Applied Entomology. Tata Mc Graw Hill Publishing Company Ltd., New Delhi.

Parveen, S., Choudhary, J.S., Thomas, A. and Ramamurthy, V.V., (2015). Biology, morphology and DNA barcodes of *Tessaratoma javanica* (Thunberg) (Hemiptera: *Tessaratomidae*). *Zootaxa* 3936(2): 261-271.

Prasad, V.G. and Singh, R.K. (1981). Prevalence and control of litchi mite, *Aceria litchi* Keifer in Bihar.*Indian Journal of Entomology* 43(1): 67-75.

Rai, M., Dey, P., Nath, V. and Das, B. (2002). Litchi: Production technology. Technical Bulletin No. 006 (NATP) (PSR-42). Horticulture & Agro-Forestry Research Programme, Ranchi.

Ray, R.R.R and Mukherjee, U. (2012). Effect of abiotic factors on the incidence of litchi leaf roller, *Dudua aprobola* Meyrick (Lepidoptera: Tortricidae) in Bihar. *Pest Management in Horticultural Ecosystems* 18(2): 210-212.

Schulte, M. J., Martin, K. and Sauerborn, J. (2007). Biology and control of the fruit borer, *Conopomorpha sinensis* Bradley on litchi (*Litchi chinensis* Sonn.) in northern Thailand. *Insect Science* 14(6): 525-529.

Sharma, K. K., Jaiswal, A. K. and Kumar, K. K. (2006). Role of lac culture in biodiversity conservation: issues at stake and conservation strategy. *Current Science* 91(7)10: 894-898.

Singh, H.S., G. Sangeetha, S.K. Srivastava, Jaipal Singh Choudhary, Kuldeep Srivastava, Bijan Kumar Das, Ajoy Kumar Sahoo, Debasis Mishra, Sanghamitra Pattnaik *et al.* (2014). Emerging pests of fruit crops, like mango, litchi, bael, tamarind, sweet orange, banana, papaya and guava in eastern India. *Journal of Applied Zoological Research* 25(2): 161-169.

Singh, H.S., Nath V., Singh, A. and Pandey, S.D. (2012). Litchi: Preventive practices and curative measures. Satish Serial Publishing House, Delhi. 479 pp.

Srivastava, K., Pandey, S.D., Patel, R.K. and Kumar, A. (2016). Litchi: Management of fruit and shoot borer. Extension folder-NRCL-EB-19. ICAR-NRC on Litchi, Muzaffarpur, India.

4

Biointensive Integrated Pest Management of Citrus

Manish Chandra Mehta, Ingle Dipak Shyamrao and M. Raghuraman

Introduction

A symposium organized by the Food and Agricultural Organization (FAO) in 1966 and International Organization of Biological Control (IOBC) in 1967, felt the need for the development of a new concept called "Integrated control" which was later replaced by "Integrated Pest Management". The origin of IPM was necessary after the serious failure of the chemical age which started in the 1940's. The pesticides were used as the only tactic to completely annihilate the pests from the crop ecosystem, which did miraculously decreased the pests and increased the production of the crop extensively, but they failed to realize the balance of life, by nature and by virtue of the selection pressure pests started developing resistance against all kind of pesticides being dumped on them. While the pests were getting resistant, the dumped pesticides were accumulating in the environment and biomagnifying inside the bodies of animal (WHO, 1990; Alewu & Nosiri, 2011; Eskenazi *et al.*, 1999). The indiscriminate use of pesticides led to the significant health hazard to the humans (WHO, 1990; Alewu & Nosiri, 2011; Sanborn *et al.*, 2007; Mnif*et al*; 2011; Semchuk *et al.*, 1992; Goad *et al.*, 2004; McKinlay *et al.*, 2008; Gasnier *et al.*, 2009; Lin *et al.*, 2015; Waddell *et al.*, 2001), reproductive malfunction in male (Jamal *et al.*, 2015), nervous system (Jaga and Dharmani, 2003; Rosenstock *et al.*, 1991; Wesseling *et al.*, 2002; Eskenazi *et al.*, 2006), cardiovascular system (Hung *et al.*, 2015) and ultimately death (WHO, 1990; Gunnell *et al.*, 2007). From the lessons acquired of the hazards, the concept of IPM proposed the concept of "pest eradication" was replaced with "pest management". There are more than 64 definitions of IPM but the most accepted one came from FAO (1967) which defined IPM as "a system in context with associated environment and the population dynamics of pest species, utilizes all suitable techniques and methods in as compatible manner as possible and maintains the pest populations at level below those causing economic injury". IPM decreased the pest pressure of the crop and increased the productivity of the crop and income of the farmers. IPM is a complex approach that advocates the integration of several components such as physical, mechanical, biological, legal, ecological and chemical tactics, in a compatible manner in a complete ecological set up, so that natural mortality has a role in checking the pest population building up above economic injury level (EIL).

Further increase in productivity and income of the farmer along with the environment protection and human health was realized with the concept of Biointensive Integrated Pest Management. It includes ecological and economical factors into biologically manipulated agricultural system. With its implementation follows many benefits such as (a) less expenditure on chemical (b) reduced risk of pesticide pollution in surface or ground water by runoff from the farms (c) reliance on synthetic pesticides decreases, which ultimately may solve the issue of resistance and resurgence among the insects (d) flourishing of the beneficial insects such as natural enemies or pollinators (e) availability of clean environment (e) the residue accumulation in the soil, and water is prevented. Biointensive Integrated Pest Management can be defined as "A system approach to pest management based on an understanding of pest ecology. It begins with steps to accurately diagnose the nature and source of pest problems, and then relies on a range of preventive tactics and biological controls to keep pest populations within acceptable limits. Reduced-risk pesticides are used if other tactics have not been adequately effective, at a last resort, and with care to minimize risks." (2). The decision and design making options of Biointensive Integrated Pest Management living within the ecological boundary, separates itself from the conventional IPM. Biointensive Integrated Pest Management mostly focuses on reducing the ecological carrying capacity for the pests. Before the application of major tactics Biointensive Integrated Pest Management includes some pre-requisites such as collection of data through survey, identification of the major fruit pest, sampling and pest forecasting, scouting the pest population density, seasonal abundance and proper record keeping. The major tactics remains almost same as that of conventional IPM with the addition of biopesticides and biorationals. Owing to the benefits of Biointensive Integrated Pest Management it was applied in various fruit, vegetable, ornamental, tuber, medicinal, aromatic, plantation, and spice crops.

Citrus belongs to the family Rutaceae. The genus citrus consists of many commercially available species in India such as grape fruit (*Citrus paradise*), lemons, limes (*Citrus aurantifolia*), sweet orange (*Citrus sinensis*), mandarin (*Citrus reticulata*), trifoliate orange (*Poncirus trifoliata*). Most of the citrus are cultivated by grafting (Kartz & Weaver, 2003). Some of the important root stock species used are citrus jambhiri lush, Citrus kama Raf. Khama Khatta (Khatta), and Citrusreshni tanaka (Cleopatra). Citrus was originated in South-

East Asia, and is mostly cultivated in tropical to temperate regions of the world and subtropical belts of North India. Citrus is cultivated in an area of 973 mha and the production is about 12253 bT (National Horticulture Board, 2019). Among them manadarin alone cover 50% of the total cultivated area with an area of 404 mha and a production of 4964 bT.The leading orange producers are the U.S.A, Brazil, Central and South America, South Africa, Japan, China, India and Mediterranean countries.Citrus has immense of nutritional quality, with the fruits packed up with thiamine (Vit B), folic acid (Ting, 1977), Vitamin C (Streiff, 1971), Vit E (Newhall & Ting, 1965). Citrus fruits have important digestive enzymes for human body. It also has minerals such as phosphorus, zinc, copper, magnesium, potassium and fibers such as cellulose, hemicellulose and pectin. Citrus fruit has many medicinal properties, manadarin is antispasmodic, sedative and helps in digestion. They help treating anorexia and are beneficial for the skin. Eating citrus fruits and drinking juice reduces the risk of cancer and heart diseases. Citrus has anti-carcinogenic, anti-bacterial, antiviral and anti-fungal property. In China, about 74 species of insects among 36 families and 9 orders attacks the citrus orchard (Yang, 2004). Insects are very destructive to the production of citrus fruit.Thus to attain marginal or above marginal production of citrus fruit, controlling the pest below EIL (Economic Injury Level) is very essential.

Biointensive Integrated Pest Management in Citrus

The major pests of the citrus are discussed along with their Biointensive Integrated Pest Management strategies.

1. Citrus aphid

- Black Aphid: *Toxapotera aurantia*
- Brown aphid: *Toxaptera citricida*

1.1 Systemic position of citrus aphid

Kingdom	:	Animalia
Phylum	:	Arthropoda
Class	:	Insecta
Order	:	Homoptera (Hemiptera)
Suborder	:	Sternorrhyncha
Superfamily	:	Aphidoidea
Family	:	Aphididae
Subfamily	:	Aphidinae
Genus	:	*Toxaptera*
Species	:	*aurantia or citricida*

1.2 Symptom of damage

Both nymph and adult suck the plant sap by piercing into the leaves with the help of stylet that results in the curling of leaves and leaves fall off immaturely and in rare situation galls are formed (Metcalf, 1962). They are vectors of citrus *tristeza* virus (CTV), which causes rapid decline and ultimately death of the citrus tree grown on sour orange root stock. During the vegetative growth period the aphids congregate on the tender leaves,young shoots and during onset of reproductive stage, targets the flower buds and cause direct damage to the plant (Hall & Ford, 1933). The virus affected citrus plant exhibits stem pitting, reduced plant vigor, reduction in quality and quantity of the produce, reduced fruit size and number on the tree.Honey dew secretion leads to the development of sooty mould fungus. Leaves turn to black in color and photosynthesis is adversely affected.

1.3 Seasonal incidence

As compared to the plains, hilly areas get more infestation of this pest (Dixon *et al.*, 1982). In hilly region bi-modal trend of population incidence is observed (Agarwala & Bhatacharya, 1995), with peak infestation in May, during summer. Usually the population density is higher during summer and autumn while lower during rainy and winter season. While in plains, the higher pest population was observed during the month of January when the temperature was lower.

1.4 Biology

Toxoptera aurantia reproduces parthenogenetically with thelytokous mode of parthenogenesis the nymphal period is 6-8 days. The nymph settles on the tender leaves, young shoots and flower buds and suck the sap. The meanpre-reproductive period of nymph is 8.1 days, with a longevity of 28.4 days. Biotic potential or fecundity accounts to 58.5 offspring/female (Yokomi, 2009). Two adult forms are visible in the colony (a) adult winged form (alate) (b) adult wingless form (apterate). Development of Winged morphs is stimulated withcrowding of nymphs/ adults in the colony or during the food shortage.

2. Asian citrus psyllid, *Diaphorina citri*

2.1. Systemic position

Kingdom	:	Animalia
Phylum	:	Arthropoda
Class	:	Insecta
Order	:	Homoptera (Hemiptera)

Suborder	:	Sternorrhyncha
Superfamily	:	Psylloidea
Family	:	Psyllidae
Genus	:	*Diaphorina*
Species	:	*citri*

2.2. Symptom of damage

Nymph and adults both are the damaging stages and the suck the phloem sap from the young shoots, leaves and buds causing death of the terminal shoot, loss of vigor, withering and wiltingof leaves and damage to the canopy. bearing high infestation. They excrete honeydew that causes sooty mould formation, and black incrustations are formed on the leaves and the shoots that causes direct damage and upto 83% - 95% of the 25% yield loss in citrus is incurred (Shivankar & Singh, 2006). They also transmit a disease called "citrus greening" (da Graca 1991; Halbert and Manjunath, 2004) or "Huanglongbing" (Tirtawidjaja *et al.,* 1965; Teaching & Research Group of Phytopathology of Guangdong Agricultural and Forest College 1977; Xu *et al.,* 1988) and the causal agent involved is a gram-negative, motile bacterium *Candudatusliberibacter* spp. (Jagoueix *et al.,* 1994; Garnier *et al.,* 2000).

2.3. Seasonal incidence

The population density of citrus psylla depends on the temperature and availability of the vegetative source. The abundance was low during May-June (Bindra, 1970), while the numbers were higher during February-March, June-July, and September-October (Kameswara, 1984). The higher population was correlated with the vegetative flush, with higher incidence during *ambebahar* followed by *mrig*and *hasta bahar*. The nymphal population shows year-round presence, except during December and January (Gupta & Bhatiya, 2000; Sharma, 2008) with peak population in the month of February, July and October (Kuchanwar *et al.,* 1985) and March-September (Bihari & Narayan, 2010).

2.4. Biology

About 180-520 eggs are laid by a female (Pande, 1971), and usually deposited on the emerging unfolded leaf or on the terminal shoot. The incubation period is about 4-18 days. The life cycle completes in about 15 to 47 days. There are 5 nymphal instars lasted for 10-20 days (Pandey, 1971). Threre were 10 overlapping generations recorded.

3. Fruit sucking moth, *Otheris maternal, O. ancilla, O. fullonica, O. salaminia,* and *O. Rhytiahypermnestra*

3.1. Systemic position

Kingdom	:	Animalia
Phylum	:	Arthropoda
Class	:	Insecta
Order	:	Lepidoptera
Family	:	Noctuidae
Genus	:	*Otheris*
Species	:	*fullonica, ansilla, maternal*

3.2. Symptoms of damage

The fruit sucking moth could be seen in the citrus orchard at dawn and they attack the fruit at night. Adults are the damaging stage while the caterpillar are defoliator of weed hosts such as *Tinospora cardifolia, Cocculus pendulus,* and *Cocculus hirsutus,* and do not attack the citrus plant. They pierce the rind and suck the juice through the long proboscis. This puncture causes the fruit to rot and later the fruit falls down. Their proboscis has dentate tips with which they pierce the fruits. Apart from citrus they also attack the fruits of guava, pomegranate, grapes, fig, sapota, papaya, mango and tomato.

3.3. Seasonal incidence

Otheris maternal start oviposition form the end of July with a peak population during the month of August to October, and later the population decreases at the end of January. *Tinospora cordifolia,* are source of breeding for the larva. Egg laying startson *T.cordifolia,* by the end of June with even more eggs during July first week (Sontakay, 1944). The oviposition by O.*materna* strated fromend of April to early May to January (Bhumannavar and Viraktamath, 2001a), which attain its peak during October to December month. Oviposition by of *O. fullonia* starts from the September first week, ending till second week of January,with peak activity during October-November. The activity of *O. fullonica* and *O. homaena* exhibited near trends.

3.4. Biology

The eggs of fruit sucking moth are whitish yellow and spherical. The dormant period of the egg is 2-3 days. The larva has 5 instars and the total larval period is 13-15 days. Pupal period lasted for 21 days. The pupa was brownish to black in color and cylindrical in shape. The adults are robust and body was covered with orange colored scales on both sexes while female moth had 3 black triangles on the forewings while only 2 faint triangles in the male (Patel & Patel, 2006).

4. Citrus Leaf Miner, *Phyllocnistis citrella*

4.1. Systemic position

Kingdom	:	Animalia
Phylum	:	Arthropoda
Class	:	Insecta
Order	:	Lepidoptera
Family	:	Gracillariidae
Genus	:	*Phyllocnistis*
Species	:	*citrella*

4.2. Symptom of damage

Caterpillar attacks the tender leaves with the help of toothed mandible and feeds on the oozed-out sap from epidermal cells leads to silvery appearance. The first three instars feeds on leaves but the fourth instar remain dormant and produces silk for pupal case. Distorted and crinkled leaves are initial symptoms of damage and severe attack causes defoliation. This encourages the problem of a disease called citrus canker, whose causative agent, *Xanthomonas axonopodis* pv. *Citri*, enters through the direct feeding sites of leaf miner.

4.3. Seasonal incidence

This pest can be seen to infesting citrus tree in two peaks, with first peak during the first week of October and second peak during the second week of October (Mane *et al.*, 2018). The abundance increases during the late spring and decreases during the early winter (Powell *et al.*, 2007). The infestation of the leaf miner starts with the new flush of citrus leaves, being lowest during spring while increasing in numbers during early and late summer until onset of winter (Legaspi *et al.*, 1999).

4.4. Biology

Dome shaped, translucent eggs are deposited singly on the midrib of underside of the leaves but during high infestation the upper surface of leaf or the stem are also chosen as oviposition site (Ware, 2014), with incubation period of 2-3 days (Kaf *et al.*, 2006). This pest has four larval instars with an average larval period of 1-2 days each. Pupal period is about 6-8 days. The adults are soft, delicate, silver brown moth with narrow forewings and fringed hindwings with long hairs. The adult male on an average life for 1-2 days and female for 2-3 days. Total life cycle takes around 16-20 days (Kaf *et al.*, 2006).

5. Citrus Butterfly, *Papilio demolleus, Papilio polytus* and *Papilio Helenus*

5.1. Symptoms of damage

The caterpillars feed voraciously on the leaves and especially young seedlings and trees are seriously affected. Complete defoliation occurs in severe attack.

5.2. Seasonal incidence

This pest occurs almost throughout the year, except during the months of January and February (Mal *et al.*, 2014).

5.3. Biology

Pale yellowish, smooth, flat, yellowish eggs are laid singly, on the abaxial surface of the young leaves and shoots. The incubation period is about 1-4 days (Alturi *et al.*, 2002). There are 5 larval instars (Singh and Gangwar (1989); Grund (2002) and Phartiyal *et al.*, 2012) with a mean duration of 2-4 days of first to fourth instars while the fifth instar duration lasted for 4-6 days. Second instar larva resembled that of a bird dropping. Third and fourth instars resembles equally with green color, while the fifth instar was cylindrical with tapered end and green in color. Total larval duration is about 14-18 days or 17-40 days (Phartiyal *et al.*, 2012) depending on the host. Pupa is greenish with dirty brown color, naked and stout, attached to the thicker stem of host (Patel *et al.*, 2017). Pupal period lasted from 6-9 days. Total life cycle covered about 23-26 days (Patel *et al.*, 2017).

6. Citrus Mite

- Citrus rust mite, *Phyllocoptruta olievera*, Eriyophidae.
- Oriental red mite, *Eutetranychus orientalis*, Tetranichidae
- Citrus red spider mite, *Panonychus citri*, Tetranichidae.
- Flat mite, *Brevipalpus phoenicis*, Tenuipalpidae.
- Two spotted spider mite, *Tetranychus urticae*, Tetranichidae.

6.1. Systemic position of two spotted spider mite

Kingdom	:	Animalia
Phylum	:	Arthropoda
Class	:	Arachnida
Subclass	:	Acari
Suborder	:	Acariformes
Order	:	Trombidiformes
Suborder	:	Prostigmata
Superfamily	:	Tetranychoidea
Family	:	Tetranychidae
Subfamily	:	Tetranychinae
Genus	:	*Tetranychus*
Species	:	*urticae*

Mite	Symptom of damage	Seasonal incidence	Biology
1. Citrus rust mite	Browning of the leaves and sickly appearance of the fruits. bronzing of the green twigs, fruit drop (Allen,1978; Huan *et al.*, 1992 and Yang *et al.*,1994). Juice volume of fruit decreases, soluble solids, acids, acetaldehyde, & ethanol increases.	---	**Egg:** spherical, pale, laid singly on fruit/leaf surface, EP-3days (summer), 5.5 days (winter), larva and nymphal stages resembles adults. NP- 3.1 (summer), 9.7 (winter) (Sarada *et al.*, 2018).Adult: elongated, wedge shaped. Total life cycle- 7-10 days (summer), 14 days (winter), (Brusssel, 1975).
2. Oriental red mite	Chlorotic appearance of infested leaves that finally becomes weak and drops.	Incidence is observed during the month of April-June.	One generation may be completed in 10 days depending on temperature with 10 to 20 generations per year.
3. Citrus red spider mite	Infestation is generally on the leaves and young to mature fruits, causes bleaching appearance, with straw yellow colored fruits Heavy infestation leads to the falling of leaves, defoliation, dieback of twigs and fruit fall.	Incidence is usually observed during the month of May.	EP: 1-2 days (April), 11-15 days (January) LP: 1day (May), 2-4 day (December) PP: 1day (May), 3-5 days (December & January), DP: 1 day (August), 10-12 days (January) (Naher*et al.*, 2008). Total life cycle – 7-12 days (summer), 12-21 days (winter)
4. Flat mite	Causes brown scabbing/leathering of fruit called "alligator skin"	Incidence was at peak during month of November but occurrence was observed throughout the year.	Elliptical, reddish orange egg, laid singly in cracks of fruit twig, and leaves. Larval and nymphal instars are present. Adults are small and flat (Elmmer&Jeppson, 1957).
5. Two spotted spider mite	Chlorotic appearance of infested leaves that finally becomes weak and drops.	Incidence was observed during the month of May-July	Spherical eggs, laid singly on underside of leaf, scattered in webbings, EP: 7 days (summer), 14-21 days (winter), LP: 1-3 days (Dhooriya,1985), PP: 1-3 days, DP: 2-3 days, larval and nymphal period combinedly takes 7-10 days. protonymph is oval, pinkish & has 4 pair of legs, deutronymph and protonymph enters the quiescense before transforming to subsequent stage. Male and females are easily distinguished as the female is wedged shaped with pointed abdomen.

Citrus mites with their biology and seasonal abundance

Biointensive Integrated Pest Management strategies for the pests of citrus

The Biointensive Integrated Pest Management strategies for the citrus pests are as follows:

Pre-requites of Biointensive Integrated Pest Management

Before heading towards the main tactics of management, the prerequisite of Biointensive Integrated Pest Management is very essential. Collection of baseline data, field scouting, sampling, monitoring, survey, pest identification and record keeping comes under the pre-requisites. For monitoring the citrus aphid, yellow water-pan traps, with water maintained at depth of 3-4 cm and added detergent is commonly used. Visual inspection also helps in monitoring. For mites, collection of leaves from fields and microscopic observation is used as a method of monitoring. In case of citrus leaf miner, the field collected leaves are observed for the presence of mines. Visual observation and stereoscopic examination for determining the density of the whitefly and the numbers of natural enemies. Examination of fruit sucking moth is done by trapping the adults in orchard during dawn. Citrus psylla are monitored using yellow sticky trap, blue sticky card traps and multi-lure traps layered with ethylene glycol or dichlorovos are also used in monitoring leaf hoppers, whiteflies. The immature stages such as eggs, larva and the feeding damage caused by the larval instars are used in monitoring the citrus butterfly. Citrus fruit fly is monitored using sex-pheromone traps, IPMT (International Pheromone Plastic McPhail trap), food attractants and mass trapping. Data recording by the farmers based on plant health, presence of pests, natural enemies, input costs, and harvest parameters such as yield, and benefit, also is helpful as pre-requisite of Biointensive Integrated Pest Management.

Biointensive Integrated Pest Management tactics

1. Cultural practices

• Cultural practices are the manipulation of the agronomical practices which favors the productivity of the crop and reduces the pest load.

• Sanitation and ploughing of the orchard wipe out the chances of carry-over of the pest's dormant stages. They are found effective in managing scales, mealy bugs and fruit sucking moth on citrus.

• Pruning is an important practice in citrus orchard to revitalize the tree, by removing the old infested twigs, that in turn creates proper space for sunlight and air. Pruning and destruction of the ant colonies in the orchard helps to control mealy bugs.

- Growing attractant plants such as cornflower (*Centaurea cyanus*), coriander attracts the natural enemies of citrus mealy bug and that of carrot family, sunflower family, alfalfa, corn, shrubs attract the natural enemies of citrus butterfly.

- Well irrigated orchard has less problem of citrus mites. Practices such as fertilization, irrigation and pruning should be well regulated to enhance discrete flush patterns over large areas, breaking generations overlap (Anagnou-Veroniki, Volakakis, & Gianoulis, 1995; Michelakis &Vacante, 1997).

- Bagging of the fruits and smoking of the citrus orchard during dawn is practiced for controlling fruit sucking moth.

2. Mechanical and physical practices

- Collection and destruction of the larva, nymph or the infested plant part is practiced for controlling citrus blackfly, mealy bug and citrus psylla, while destruction of the alternate host is carried out in case of fruit sucking moth.

- Use of pheromone trap @ 5/acre for managing citrus leaf miner and yellow sticky trap for aphids. Hand picking of the larva and killing is practiced against citrus butterfly.

- Collection and destruction of rotten fruits, installation of fame torch and light trap before 1 month of fruit maturation is practiced for fruit sucking moth.

- Installation of sticky bands on the tree trunk to avoid climbing by crawlers of scales and whiteflies.

3. Biological control

- Conservation of natural enemies (parasites, predators and parasitoids) helps controlling many pests of citrus.

- An exotic parasitoid *Laptomastix dactylopii* @ 5000-7000/adult/ha, is used against citrus mealy bug.

- Release of coccinellid beetles such as *Menochilus sexmaculatus* @50/ tree helps to suppress aphid population. Other coccinellid, *Cloneda sanguinea* (L.) and *Harmonia axyridis* Pallas are found effecting in controlling aphids.

- *Chrysoperla carnea*, coccinellid beetles and syrphids against citrus psylla. Conservation of *Chilocorusnigrita,* and *Scymnus quadrillum*, for controlling citrus blackfly.

- *Crytolaemes montrouzieri* should be released @10 beetles/plant against citrus mealybug. Release of *Chrysoperla bugs*@ 10-15/plant against citrus mites.

- *Tamarixia radiata* and *Diaphorencyrtus aligarhensis* (Nymphal and adult parasitoid) are conserved against citrus psylla.

- *Encarsia Formosa* and *Eretmocerus* spp. Release of *Trichogramma evanescens*, *Tilonomus* spp. (egg parasitoid), *Brachmeria* sp. (larval parasitoid) and *Pterolus* sp. (pupal parasitoid), controls citrus butterfly.

- Spraying of entomogenous fungus, *Bacillus thuringiensis* Berliner, nematode DD-136 strain is beneficial in controlling citrus butterfly larval stage.

- *Cladosporium oxysporum*@ 1.5×10^8 spores/ml, is effective against citrus mealybug (Samways and Grech, 1986).

- Isolates of *Metarhizium anisopliae* and *Beauveria bassiana* @ 5.29×10^5 and 4.25×10^6 conidia / ml, respectively against whiteflies, and mealybugs.

- Increased mortality of adult psyllidsis incurred by *Beuveria bassiana* and *Isariafumosoro-sea* @10^8conidia/ml (Gandarilla-Pacheco *et al.*, 2013).

- *B. bassiana*– based mycoinsecticde, and *Lecanicillium lecanii* has effective control over brown citrus aphid (Balfour and Khan, 2012)

- *Hirsutella thompsonii* mycelial formulation works effectively against rust mite infestation formulated as Mycar (Abbott Laboratories) (McCoy *et al.*, 1971).

- *Hirsutella thompsonii* is also effective *Tetranychus urticae*.

- *B. bassiana* has a significant drastic effect on the two-spotted spider mite oviposition and eggs. It is also effective against citrus flat mite (Rossi-Zalaf and Alves, 2006).

- *Heterorhabditis zealandica* and *Stinirnema yirgalemense* are highly potent to adults of mealybugs (van Niekerk and Malan, 2014).

- *Bacillus thuringiensis*, is the only bacterial pathogen that is highly effective against citrus pests, especially lepidopterans (Lacey and Shapiro-Ilan, 2008). Although *Serratia entomophila*is also used and effective against citrus leafminer, *Phyllocnistis citrella* (Meca *et al.*, 2009) and *Bacillus subtilis* and *B. mycoides* are also found to infect leaf miner (Saiah *et al.*, 2011).

4. Botanicals

- Tobacco extract @ 2%, neem seed extract @ 2%, and garlic extract @ 2% are used againt the citrus aphid, among which tobacco extract and neem extract sprayed on the new flushes of leaves gave much control against the *Toxaptera aurantia* and *Toxapteracitricida*. (Sohail *et al.*, 2012).

- Pepper extract (Tondexir) 2 ml/l, garlic extract (insecticide emulsion) @ 4 ml/l, are effective against citrus leaf miner.

- Neem seed extract reduced the longevity of the adults and nymphs. Disturbs molting and causes increased mortality in brown citrus aphid (Tang *et al.*, 2002).

- Aqueous garlic bulb extract, neem extract, castor extract, ocimum extract are proved to have effective antifeedant, deterrent, and ovicidal property against citrus butterfly (Elayidam, 2014).

5. Chemical control

- Spraying of 2.5 lt. of ethion 50 EC or 3.125 lt. of Triazophos 40 EC in 1250 liters of water/ha during April-May and September-October against citrus whitefly.

- Spray of 400 g of thiamethoxam 25 WG or 500 ml of imidachloprid 17.8 SL or 3.125 lt. of dimethoate 30 EC or 1.5 lt of monocrotophos 36 SL or 2.5 lt. of Oxydementon methyl 25 EC in 1250 lt. of water/ha during March for controlling citrus psylla.

- Spray of 1.5 lt. of quinolphos 25 EC or 2 kg of carbaryl 50 WP in 1250 lt. of water/ha during April and October to control citrus butterfly.

- Spray 400 g of thiamethoxam 25 WG or 500 ml of imidachloprid 17.8 SL or 1.25 lt of fenvalerate 20 EC or 2.5 lt. of cypermethrin 10 EC in 1250 lt. of water/ha during April-May & August-September, to control citrus leaf miner.

- Spray tree with 2.5 kg of carbaryl 50 WP in 500 lt. of water/ha during fruit maturity against fruit sucking moth.

- Aphids can be controlled in the same manner as the citrus psylla. Citrus mite can be controlled by spraying 2.5 lt of ethion 50 EC or 1.875 lt. of Fenazaquin 10 EC in 1250 lt of water/ha on the underside of the leaves.

References

Agarwala, B.K. and Bhatacharya, S. (1995). Seasonal abundance of black aphid *Toxaptera aurantia*, in North-East India: Role of temperature. *Proceeding of Indian National Academy*. B 61, No. 5, pp. 377-382.

Alewu, B. and Nosiri, C. (2011). Pesticides and human health. In: Stoytcheva M, editor. *Pesticides in the Modern World – Effects of Pesticides Exposure.* p. 231-50.

Alturi JB, Ramona SPV, Rudi CS. *India. J Natl. Taiwan Mus.* 55: 27-32.

Anagnou-Veroniki, M. (1995). First record of citrus leafminer, Phyllocnistiscitrella (Stainton) on citrus groves of mainland and island Greece. *Annales de l'InstitutPhytopathologiqueBenaki,* 17: 149-152.

Balfour, A., Khan, A. (2012). Effects of *Verticillium lecanii* (Zimm.) Viegas on *Toxopteracitricida*Kirkaldy (Homoptera: *Aphididae*) and its para-sitoid *Lysiphlebustestaceipes*Cresson (Hymenoptera: *Braconidae*). *Plant Prot. Sci.* 48: 123-130.

Bhumannavar, B. S. and Viraktamath, C.A. (2001a). Biology, adult feeding, oviposition preference and seasonal incidence of *Othreismaterna* (L.) (Lepidoptera: *Noctuidae*). *Entomon,* 27(1): 63-77.

Bihari, M. and Narayan, S. (2010). Seasonal incidence and damage of *D. citri* kuw.on citrus. *Ann. Pl. Protec.Sci.,* 18: 31-33.

Bindra, O.S. (1970). Citrus decline in India – causes and control. Joint Publ. of P.A.U. and OSU/USAIO, Ludhiana, Punjab. pp. 64-78.

Brussel EW Van. Inter-relations between citrus rust mite, fungi, *Hirsutella thompsonii* and greasy spot on citrus in Surinam. Centre for Agricultural Publishing and Documentation Wageningen, 1975.

Da Graca JV. (1991). Citrus greening disease. *Annual Review of Phytopathology* 29(1): 109-136.

Elayidam, U.G. (2014). Efficacy of aqueous extracts of selected plant products against Papiliopolytes L. *Current Biotica* 8(3): 303-308.

Eskenazi, B., Harley, K., Bradman, A., Fenster, L., Wolff, M. and Engel, S. (2006). In utero pesticide exposure and neurodevelopment in three NIEHS/environmental agency children's center birth cohorts. *Epidemiology* 17: S103.

Eskenazi, B., Marks, A.R., Bradman, A., Fenster, L., Johnson, C. and Barr, D.B. (2006). In utero exposure to dichlorodiphenyltrichloroethane (DDT) and dichlorodiphenyldichloroethylene (DDE) and neurodevelopment among young Mexican American children.*Pediatrics* 118(1): 23-41.

Gandarilla-Pacheco, F.L., Galán-Wong, L.J., López-Arroyo, J.I., Rodrí-guez-Guerra, R., Quintero-Zapata, I. (2013). Optimization of pathoge-nicity tests for selection of native isolates of entomopathogenic fungi isolated from citrus growing areas of México on adults of *Diaphorina citri* Kuwayama (Hemiptera: Liviidae). *Fla. Entomol.* 96: 187-195.

Garnsey, S.M., Gottwald, T.R. and Yokomi R.K. (1996). Control strategies for citrus tristeza virus, In: A. Hadidi, R. K. Khetarpal, and H. Koganezawa (eds.). Plant Viral Disease Control: Principles and Practices. *APS Press.* St. Paul, MN: 639-658.

Gasnier, C., Dumont, C., Benachour, N., Clair, E., Chagnon, M.C. and Séralini, G.E. (2009). Glyphosate-based herbicides are toxic and endocrine disruptors in human cell lines. *Toxicology* 262(3): 184-91.

Goad, E.R., Goad, J.T., Atieh, B.H. and Gupta, R.C. (2004). Carbofuran-induced endocrine disruption in adult male rats. *Toxicol Mech Methods* 14: 233–9.

Gunnell, D, Eddleston, M., Phillips, M.R. and Konradsen, F. (2007). The global distribution of fatal pesticide self-poisoning: systematic review. *BMC Public Health* 7(1): 357.

Gupta, D. and Bhatia, R. (2000). Population dynamics of citrus psylla in lower hills of Himachal Pradesh. *Pest Mgmt. Eco. Zool.,* 8(1): 41-45.

Hall, W.J. and Ford, W.K. (1933). Notes on some citrus insects of southern Rhodesia. British South Africa Co., Mazce Citrus Expt. Station Publ. 2a.

Halbert, S.E. and Manjunath, K.L. (2004). Asian citrus psyllids (Sternorrhyncha: Psyllidae) and greening disease of citrus: a literature review and assessment of risk in Florida. *Florida Entomologist* 87(3): 330-353.

Hung, D.Z., Yang, H.J., Li, Y.F., Lin, C.L., Chang, S.Y. and Sung, F.V. (2015). The long-term effects of organophosphates poisoning as a risk factor of CVDs: a nationwide population-based cohort study. *PLoS One,* 10(9): e0137632.

Jaga, K. and Dharmani, C. (2003). Sources of exposure to and public health implications of organophosphate pesticides. *Rev PanamSalud Publica* 14(3): 171-85.

Jagoueix, S., Bove´, J.M. and Garnier, M. (1994). The phloem-limited bacterium of greening disease of citrus is a member of a subdivision of the proteobacteria. *International Journal of Systematic Bacteriology* 44(3): 379-386.

Jamal, F., Haque, Q.S., Singh, S. and Rastogi, S. (2015). The influence of organophosphate and carbamate on sperm chromatin and reproductive hormones among pesticide sprayers. *Toxicol Ind Health* 32(8): 1527-1536.

Kaf, A. N., Aslan L. and Ahmed I. (2006). Morphology and Biology of Citrus Leaf Miner *Phyllocnistiscitrella* (Lepidoptera: Gracillaridae) in Syria. *Arab J. Pl. Prot.* 24: 45-48.

Kameswara, R.P. (1984). Management of citrus psylla. National seminar on pest management in citrus, cotton, sugarcane and sorghum – Progress and Problems, held at Nagpur. pp. 5-7.

Katz, S. H., and W. W. Weaver. (2003). *Encyclopedia of Food and Culture*. New York: Schribner. ISBN 0684805685.

Kuchanwar, D. D., Hardas, M.G. and Sharnag at, B.K. (1985). Agrotechniques in management of citrus pests. *Pestology.*, IX(1): 5-9.

Lacey, L.A., Shapiro-Ilan, D.I. (2008). Microbial control of insect pests in temperate orchard systems: potential for incorporation into IPM. Annu. Rev. *Entomol* 53: 121-144.

Legaspi, J.C. and French J.V. (2002). The citrus leafminer and its natural enemies.http://primere.tamu.edu/kcchome/pubs/leafminer.htm.

Li, D., Huang, Q., Lu, M., Zhang, L., Yang, Z. and Zong, M. (2015). The organophosphate insecticide chlorpyrifos confers its genotoxic effects by inducing DNA damage and cell apoptosis. *Chemosphere* 135: 387-93.

Mal J., Patel M.M., Meena B.M., Nagar R., Kumawat S.C., Chauhan S.K. and Meena A. (2014). *Agriculture for Sustainable Development.* 2(1): 27-31.

McCoy, C.W., Selhime, A.G., Kanavel, R.F., Hill, A.J. (1971). Suppression of citrus rust mite populations with application of fragmented mycelia of *Hirsutella thompsonii. J. Invertebr. Pathol.* 17: 270-276.

Mane, S. B., Nagar Sasya and Kolhe P. S. (2018). Seasonal Incidence of Citrus Leaf Miner (*Phyllocnistiscitrella*Stainton) in Trans Yamuna Region of Allahabad, India.*Int. J. Pure App. Biosci.* 6(3): 726-728.

McKinlay, R., Plant, J.A., Bell, J.N.B. and Voulvoulis, N. (2008). Endocrine disrupting pesticides: implications for risk assessment. *Environ Int,* 34: 168–83.

Meca, A., Sepúlveda, B., Ogoña, J.C., Grados, N., Moret, A., Morgan, M. and Tume, P. (2009). *In vitro* pathogenicity of northern Peru native bacteria on *Phyllocnistiscitrella* Stainton (Gracillariidae: Phyllocnistinae), on predator insects (*Hippodamia convergens* and *Chrisoperna externa*), on *Citrus aurantifolia* Swingle and white rats. *Span. J. Agric. Res*. 7: 137-145.

Metcalf C.L., Flint W.P. (1962). Destructive and Useful Insects, their Habits and Control. McGraw-Hill Book Company, New York, San Francisco, Toronto, London, 1087 pp.

Michelakis, S. E., & Vacante, V. (1997). The citrus leafminer status in Greece. Integrated control in citrus fruit crops. *Bulletin-OILB-SROP,* 20: 81-82.

Mnif, W., Hassine, A.I.H., Bouaziz, A., Bartegi, A., Thomas, O. and Roig, B. (2011). Effect of endocrine disruptor pesticides: a review. *Int J Environ Res Public Health* 8: 2265-2203.

Newhall, W.F. and Ting, S.V. (1965). Isolation and identification of β-tocopherol, a vitamin E factor from orange flavedo.*J. Agric. Fd. Chem.* 13: 281-282.

Pande, Y.D. (1971). Biology of citrus psylla, *Diaphorina citri* Kuw. (Hemiptera: Psyllidae). *Israel Journal of Entomology*, 6(2): 307-311.

Patel, M.M. and Patel, R.K. (2006). Biology of Fruit Sucking Moth, *Othreis Materna* L. On Sweet Orange. *Internat. J. Agric. Sci.* 2(1): 118-121.

Phartiyal, T., Srivastava, P., Khan, M.S. and Srivastava, R.M. (2012). *J Ent. Res.* 36(3): 255-258.

Powell C.A., Burton M.S., Pelosi Robert, Ritenour M.A., and Bullock R.C. (2007). *Hortscience* 42(7): 1636-1638.

Rosenstock, L., Keifer, M., Daniell, W.E., McConnell, R. and Claypoole, K. (1991). Chronic central nervous system effects of acute organophosphate pesticide intoxica-tion.*Lancet* 338(8761): 223-7.

Rossi-Zalaf, L.S., Alves, S.B. (2006). Susceptibility of *Brevipalpusphoeni-cis*to entomopathogenic fungi. *Exp. Appl. Acarol.* 40: 37-47.

Saiah, F., Bendahmane, B.S., Benkada, M.Y. (2011). Isolement et identification de bactériesentomopathogènes à partir de *Phyllocnis-tiscitrella* Stainton 1856 dans l'Ouestalgérien. *Entomol. Faun.* 63: 121-123.

Samways, M.J., Grech, N.M. (1986). Assessment of the fungus *Clado-sporium oxysporum*(Berk. and Curt.) as a potential biocontrol agent against certain Homoptera. *Agric. Ecosyst. Environ.* 15: 231-239.

Sanborn, M., Kerr, K.J., Sanin, L.H., Cole, D.C., Bassil, K.L. and Vakil, C. (2007). Non-cancer health effects of pesticides. Systematic review and implications for family doctors.*Can Fam Physician,* 53(17): 12-20.

Singh, Y.P., Gangwar S.K. and Andaman, J. (1989). *Sci. Assoc.* 5(2): 151-153.

Shivankar, V. J. and Rao, C. N., Shyam, S. and Singh, S. (2000). Studies on citrus psylla *Diaphorina citri* Kuwayama. *Agric. Rev.*, 21(3): 199-204.

Sohail, A., Hamid, F. S., Waheed, A., Ahmed, N., Aslam, N., Zaman, Q., Ahmed, F. and Islam, S. (2012). Efficacy of different botanical Materials against aphid *toxopteraaurantii*on tea (*Camellia sinensis* L.) cuttings under high shade nursery. *J. Mater. Environ. Sci.* 3(6): 1065-1070.

Sontakay, K.R. (1944). Short note on the life and seasonal history of *Othreismaterna* L. and *Othreis fullonica* L. *Indian Journal of Entomology,* 5: 247-248.

Tang, Y. Q., A. A., and Weathersbee III, and R. T. Mayer. (2001). Effect of Neem Extract on the Brown Citrus Aphid, Toxopteracitricida and its Parasitoid, Lysiphlebustestaceipes. Subtropical Insects *Research Unit U.S. Horticultural Research Laboratorhttp://www.ars.usda.gov/main/ site_main.htm? docid=3631&modecode=66-18 05-10&page=7y 2001* South Rock Road Fort Pierce, FL 34945 USA.

Teaching and Research Group of Phytopathology of Guangdong Agricultural and Forest College. (1977). Preliminary report on transmission of citrus huanglungbin by psylla. *Guangdong Agricultural Science* 6: 50–51, 53 (in Chinese).

Tirtawidjaja, S., Hadiwidjaja, T, and Lasheen A.M. (1965). Citrus veinphloem degeneration virus, a possible cause of citrus chlorosis in Java.*Proceedings of the American Society of Horticultural Science* 86: 235–243.

Ting, S.V. (1977). Nutrient labeling of citrus products. In Citrus Science and Technology, (S. Nagy, S. Shaw and Veldhuise, Eds.) *AVI Publishers*, 2: 401-444.

Van Niekerk, S., Malan, A.P. (2014). Evaluating the efficacy of a polymer-surfactant formulation to improve control of the citrus mealybug, Planococcus citri (Risso) (Hemiptera: Pseudococcidae), using ento-mopathogenic nematodes under simulated natural conditions. *Afr. Plant Prot* 17: 1-8.

Wesseling, C., Keifer, M., Ahlbom, A., McConnell, R., Moon, J.D. and Rosenstock, L. (2002). Long-term neurobehavioral effects of mild poisonings with organophos-phate and n-methyl carbamate pesticides among banana workers. *Int J Occup Environ Health* 8(1): 27-34.

World Health Organization. (1990). *Public Health Impact of Pesticides Used in Agriculture*. England: World Health Organization.

Yang, Y., huang, M., Andrew, Beattie G. Yulu c., Ouyang, G, and xiong, J. (2006). Distribution, biology, ecology and control of the psyllid *Diaphorina citri* Kuwayama, a major pest of citrus: A status report for China. *International Journal of Pest Management*, 52(4): 343-352.

5

Biointensive Integrated Pest Management of Banana

Fouzia Bari and S.K.Senapati

Introduction

Banana (*Musa* sp; family: Musaceae) is the second most important fruit crop in India next to mango. Its year round availability, affordability, varietal range, taste, nutritive and medicinal value makes it the favorite fruit among all classes of people. It has also good export potential. Banana evolved in the humid tropical regions of South East Asia with India as one of its center of origin. Modern edible varieties have evolved from the two species – *Musa acuminata* and *Musa balbisiana* and their natural hybrids, originally found in the rain forests of S.E. Asia. During the seventh century AD its cultivation spread to Egypt and Africa. Banana and plantains are grown in about 120 countries. Total annual world production is estimated at 86 million tonnes of fruits. India leads the world in banana production with an annual output of about 14.2 million tonnes. Other leading producers are Brazil, Eucador, China, Phillipines, Indonesia, Costarica, Mexico, Thailand and Colombia. In India banana ranks first in production and third in area among fruit crops. It accounts for 13% of the total area and 33% of the production of fruits. Production is highest in Maharashtra (3924.1 thousand tones) followed by Tamil Nadu (3543.8 thousand tonnes). Within India, Maharashtra has the highest productivity of 65.70 metric tones /ha. against national average of 30.5 tonnes/ha. The other major banana producing states are Karnataka, Gujarat, Andhra Pradesh and Assam.

Economic Importance of Banana

Banana is a very popular fruit due to its low price and high nutritive value. It is consumed in fresh or cooked form both as ripe and raw fruit. Banana is a rich source of carbohydrate and is rich in vitamins particularly vitamin B. It is also a good source of potassium, phosphorus, calcium and magnesium. The fruit is easy to digest, free from fat and cholesterol. Banana powder is used as the first baby food. It helps in reducing risk of heart diseases when used regularly and is recommended for patients suffering from high blood pressure, arthritis, ulcer, gastroenteritis and kidney disorders. Processed products, such as chips, banana puree, jam, jelly, juice, wine and halwa can be made from the fruit. The tender stem, which bears the inflorescence is extracted by removing the leaf sheaths of the harvested pseudostem and used as vegetable. Plantains or cooking bananas are rich in starch and have a chemical composition similar to that of potato. Banana fibre is used to make items like bags, pots and wall hangers. Rope and good quality paper can be prepared from banana waste. Banana leaves are used as healthy and hygienic eating plates. In India, about 19 insect pests have been found associated with banana from planting to harvesting (Padmanaban *et al.*, 2002). Singh (1970) and Basak *et al.* (2015) detailed the major banana pests of India.

List of Insect Pest Damaging Banana

1. Banana rhizome weevil: *Cosmopolitus sordidus* Germar (Coleoptera: *Curculionidae*)

2. Banana stem weevil: *Odoiporus longicollis* Olivier (Coleoptera: *Curculionidae*)

3. Banana leaf eating caterpillar: *Spodoptera litura* Fabricius (Lepidoptera: *Noctuidae*)

4. Banana aphid: *Pentalonia nigronervosa* Coquerel (Hemiptera: *Aphididae*)

5. Banana thrips:

   • Rust thrips: *Cheatanophothrips signipennis* Bagnall (Thysanoptera: *Thripidae*)

   • Leaf thrips: *Helionothrips kadaliphilus* Ramak (Thysanoptera: *Thripidae*)

   • Flower thrips: *Thrips florum* Schumtz (Thysanoptera: *Thripidae*)

6. Banana leaf and fruit scarring beetle: *Nodostoma (Basilepta) subcostatum* Jacoby. (Coleoptera: *Chrysomelidae*)

7. Banana lacewing bug: *Stephanitis typicus* Distant (Hemiptera: *Tingidae*)

1. Banana rhizome weevil: *Cosmopolitus sordidus* Germar (Coleoptera: *Curculionidae*)

• The banana weevil *Cosmopolites sordidus* (Germar) is the most destructive insect pest of *Musa*.

• It is a serious production constraint and a principal factor contributing to the decline and disappearance of banana.

Biology and nature of damage

- The female weevil lays its eggs in the corm and pseudostem base of the banana plant.
- The emerging larvae bore through the corm.
- The resulting larval damage impedes water and nutrient uptake and reduces the stability of the plant.
- Weevil attack can interfere with crop establishment, reduce bunch size, and lead to plant loss, mat die-out and shortened plantation life.

Biointensive Management

- Field should be cleaned by removing the dried leaves and plant debris from the field and destroy it by burning or by dumping of leaves in mulching pit and covering with soil.
- Use healthy, uninfected sucker or rhizomes for planting time.
- Removal of infested pseudostems below ground level
- Trimming the rhizome is important operation to minimize this pest
- Prune the side suckers at monthly interval.
- Botanicals: 5 % aqueous leaf extract, 3% oil emulsion spray of neem seed extract of orange and leaf extract and also leaf extract of lemon grass are found effective in controlling banana rhizome weevil.
- Regular monitoring of weevil by using banana longitudinal cut stem trap of 30 cm size @ 10-15 per acre.
- In case once weevil is attracted to the laid traps, place longitudinal split banana traps @ 100 ha with bio control agents like entomopathogenic fungus *Beauveria bassiana* or entomopathogenic nematode, *Heterorhabditis indica* @ 20 g/trap.
- These bio control agents have to be swabbed on the cut surface of the stem traps and keep the cut surface facing the ground.
- Pheromone (cosmolure) trap @ 5 traps / ha and the position of traps should be changed once in a month.

2. Banana stem weevil: *Odoiporus longicollis* Olivier (Coleoptera: *Curculionidae*)

- Banana stem weevil is the serious pest of banana and damaging the stem of the banana plant, there by leads to enormous loss in yield as well as productivity.
- Pest is present throughout the year.

Biology and nature of damage

- Adults are either reddish brown or black.

- The pest is active during summer and monsoon months.

- The grubs bore into the stem and feed within the stem.

- An initial symptom is in the form of exudation of plant sap and blackened mass that comes out from the hole bore by the grub.

- Finally the whole plant dies.

- Grub bore into pseudostem making tunnels.

- Tunneled part decomposes and pseudostembecomes weak.

- They also cut holes on outer surface later blackened mass comes out from the borehole and leads towilting of the plant.

Biointensive Management

- Field should be cleaned after removing the dried leaves and plant debris from the field and destroy it by burning or by dumping of leaves in mulching pit and covering with soil.

- Use healthy, uninfected sucker or rhizomes for planting time.

- Prune the side suckers at monthly interval

- Use healthy and insect free suckers

- Do not dump infested materials into manure pit

- Uproot infested trees, chop into pieces and burn

- Regular monitoring of weevil by keeping banana traps viz. Longitudinal cut stem trap of 30 cm size @ 10-15 per acre.

- In case once weevil is attracted to the laid traps, place longitudinal split banana traps @ 100 ha with bio control agents like entomopathogenic fungus *Beauveria bassiana* or @ 20 g/trap. These bio control agents have to be swabbed on the cut surface of the stem traps and keep the cut surface facing the ground.

- Keep pheromone (cosmolure) trap @ 5 traps / ha. The position of traps should be changed once in a month.

- Botanicals: Neem oil spray of 2% as a whorl application, 5 % aqueous leaf extract of neem and NSKE 5% and 10 % are found effective in controlling banana stem weevil.

3. Banana leaf eating caterpiller: *Spodoptera litura* Fabricius (Lepidoptera: *Noctuidae*)

- It is found throughout the tropical and sub tropical parts of the world, wide spread in India.
- Besides tobacco, it feeds on cotton, castor, groundnut, tomato, cabbage and various other cruciferous crops.

Biology and nature of damage

- Female lays about 300 eggs in clusters.
- The eggs are covered over by brown hairs and they hatch in about 3-5 days.
- Caterpillar measures 35-40 mm in length, when full grown. It is velvety, black with yellowish green dorsal stripes and lateral white bands with incomplete ring like dark band on anterior and posterior end of the body.
- It passes through 6 instars. Larval stage lasts 15-30 days.
- Pupation takes place inside the soil, pupal stage lasts 7-15 days.
- Moth is medium sized and stout bodied with forewings pale grey to dark brown in colour having wavy white criss cross markings.
- Hind wings are whitish with brown patches along the margin of wing.
- Pest breeds throughout the year. Moths are active at night.
- Adults live for 7-10 days.
- Total life cycle takes 32-60 days.
- Are eight generations in a year.
- Young larvae feed by scrapping the leaves from ventral surface.
- Later on feed voraciously at night on the foliage.

Biointensive Management

- Deep summer ploughing help to expose the pupae for predators.
- Hand picking and mechanical destruction of egg masses and caterpillars in early stage of attack.
- Use healthy, uninfected planting material.
- Remove and destroy the rolled leaves with larvae and pupae of leaf rollers.
- Grow repellant plants like Ocimum/basil with banana

- Use pheromone trap@ 10-12 traps/ha. The position of traps should be changed once in a month.
- Use light trap to attract and kill the adults.
- Field release of egg parasitoids such as *Telenomus spodopterae, T. remus*
- Encourage the activity of larval parasitoids *Ichneumon promissorius, Carcelia* spp., *Campoletis chlorideae*
- For a standing crop also apply 2 kg of farmyard manure enriched with bio-pesticide. *T. harzianum* and *P. lilacinus* to be repeated 6 monthly intervals.
- Azadiractin 0.15% spray @5ml/lit of water is found effective in controlling leaf eating caterpillar.

4. Banana aphid: *Pentalonia nigronervosa* Coquerel (Hemiptera: Aphididae).

- It is found in young foliage and mostly damage the soft tissue and early plants.

Biology and nature of damage

- Oval or slightly elongated, reddish brown with six segmented antennae. Small to medium sized aphids, shiny, reddish to dark brown or almost black.
- They have six segmented antennae and prominent dark veins.
- Adults start producing young one day after reaching maturity.
- They can give birth to 4 aphids per day with an average production of 14 off spring per female.
- Leaves are bunched into a rosette appearance.
- Leaf margins are wavy and upward rolling.
- Do not produce bunches.
- It is vector of bunchy top disease and seen in colonies on leaf axils and pseudostem.

Biointensive Management

- Use yellow sticky trap @ 10-12/ha
- Clean cultivation should be followed
- Encourage activity of predator coccinellids such as *Scymnus, Chilomenes sexmaculatus*, and lacewing, *Chrysoperla zastrowi sillemi*

- Rogue out the virus affected plants before spraying
- Use healthy, uninfected planting material.
- Remove and destroy the aphid attacked leaves.
- Neem oil spray 2% @ 20 ml whorl application per banana plant.

5. Banana thrips: They are present throughout the year.

- Rust thrips: *Cheatanophothrips signipennis* Bagnall (Thysanoptera: Thripidae)
- Leaf thrips: *Helionothrips kadaliphilus* Ramak (Thysanoptera: Thripidae)

Biology and nature of damage

- Eggs are not visible to the naked eye. Eggs are laid just below the fruit or pseudostem surface. In summer, eggs hatch in about eight days
- The wingless creamy white larvae are smaller but have the same shape as the adult.
- The larval period lasts about 8-10 days.
- Pupae are white, 1 mm in length, similar to the larvae and can crawl.
- The pupal stage lasts 7-10 days.
- The adult is slender, 1.5 mm long, creamy yellow to golden brown with delicate feathery wings.
- The early symptoms appear as water-soaked smoky areas where the colonies congregate to feed and oviposit between touching or adjacent fruit.
- These areas then develop the typical rusty-red to dark brown-black discolouration.

Biointensive Management

- Collect and destroy the damaged leaves, flowers and fruits along with all life stages.
- Use blue pan water sticky trap @ 4-5/acre.
- Destroy all volunteer plants and old neglected plantations.
- Use healthy and pest free suckers for planting.
- Hot water treatment of suckers prior to planting.
- Bunch covers (which cover the full length of the bunch) protection applied very early.
- Regular checking of fruit under the bunch covers is essential to ensure that damage.

- Conserve predators such as ladybugs, lacewings, Syrphid flies, parasitic wasps.
- Azadiractin 0.15% spray @ 5 ml/ lit of water, Neem oil spray 2% @ 20 ml whorl application per banana plant.

6. Banana leaf and fruit scarring beetle: *Nodostoma (Basilepta) subcostatum* Jacoby (Coleoptera: *Chrysomelidae*)

- This pest is present through out the year and causes the loss in market value of the fruit by making scars.
- It causes extensive damage to leaves as well as fruits during summer and kharif season (Singh *et al.*, 1997).
- The extent of damage has been reported to be approximately 30 per cent of the banana bunches during rainy season in Bihar (Ahmad *et al.*, 2003; Mukherjee, 2004; Samui *et al.*, 2004 and Mukherjee, 2006).

Biology and nature of damage

- Adults have brown forewings with characteristics rows of small parallel dots.
- They are very good fliers. Female lay pale lemon yellow eggs, singly or in clusters varying in number from 5-45.
- Egg laying is taking place in cavities gnawed in leaf sheaths near the crown or in natural depressions that expose the surface of roots.
- After 7 to 9 days the newly hatched larvae start to feed on the young roots or to bore tunnels into the soft epidermal tissues of the older roots to feed on them.
- They have a whitish slender and hairy body and the head somewhat amber-colored.
- The pupae is dirty yellow, becoming darker as the adult becomes ready to emerge.

- This pest is associated with the plant from the sucker stage till fruiting.
- Only young leaves and fruits are attacked by this beetle.
- The beetle feeds on their surface superficially in irregular patches.

Biointensive Management

i) Botanicals: Azadiractin 0.15%@5ml/lit of water to whorl application to banana unopened leaves. Neem oil spray of 2% per banana plant.

ii) Cultural: Proper field sanitation, clean cultivation and removal of weeds from field @15 days interval.

iii) Microbial: Spray of *Beauveria bessiana* 5g/lit. Repeat spray at fortnightly interval if required and Stop spraying 15 days before bunch harvesting.

iv) Natural enemies: Adult of grass hopper control egg and larval stage of *Nodostoma subcostatum*. Red ant, preying mantid and spider are naturally present in field controls the larval and adult stage of *Nodostoma subcostatum*.

v) Practice clean cultivation by removing the grass weeds from the banana plantations.

7. Banana lacewing bug: *Stephanitis typicus* Distant (Hemiptera: *Tingidae*)

Biology and nature of damage

- Small, whitish lacewing bug found in colonies on the foliage.
- Causing a sickly and spotted appearance of the plant.
- The foliage of infested plants turns pale or yellow and dries up.
- The pest infestation is more common during the post monsoon period especially in drier regions of the country.

Biointensive Management

- Collect and destroy the damaged leaves, flowers and fruits along with life stages
- 5 % aqueous leaf extract of king bitters (*Andrographis paniculata*), 3% oil emulsion spray of neem seed extract of orange and leaf extract and also leaf extract of lemon grass are found effective in controlling banana lace wing bug.
- Application of *Beaveria bessiana* and *Hexamermis sps* @500 ijs/ml is effective for controlling banana lace wing bug.
- Use of natural enemies like *Cryptoleamus montrouzieri and Tetrastichus schoenobii* are effective in controlling bugs population.

References

Ahmad, M.A., Singh, P. P. and Singh, B. (2003). Efficacy of certain synthetic insecticides andplant products used as foliar and whorl application against the scarring beetle (*Nodostoma subcostatum* Jacoby) on banana. *Journal of Enomological Research,* 27(4): 325-328.

Anonymous, (2011). March. *Indian Horticulture Database*, www.nhb.gov.in.

Bauri, F. K, De, A, Misra, D. K., Bandyopadhyay, B., Debnath, S, Sarkar, S.K. and Avani, P. (2014). Improving yield and quality of banana cv. Martaman (Musa AAB, Silk) through micronutrient and growth regulator application. *Journal of Crop Weed,* 10: 316-19.

Mukherjee, U. (2006). Evaluation of insecticides and some eco-friendly approaches to manage scarring beetle, *Basilepta subcostatum* in banana.*Journal of Applied Zoological Research,* 17(1): 54-56.

Padmanaban, B., Rajeswari, R. and Satiamoorthy, S. (2002). Integrated management of insect pests on banana and plantation. Souvenir *Global Conference on Banana and Plantation,* 6: 28-31.

Sah, S.B., Parasnath, Prakash, S. and Kumar, Rajesh (2018). Occurrence of Leaf and Fruit Scarring Beetle, (*Basilepta* sp., *Colaspis* sp.) on Banana in Koshi Region of Bihar. *Int. J. Curr. Microbiol. App. Sci* (2018) Special Issue-7: 2778-2784.

Samui,G., Maiti, B. and Bandopadhyay, B. (2004). Field evaluation of some insecticidal treatments to control banana scarring beetle (*Nodostoma viridipenne* Motsch.). Proc. *Natl. Seminar Banana Industry – Present Scenario and Future Strategies* held at BCKV, Kalyani, 11th to 13th June, 2004: 194-202.

Singh, H.S. and. Sangeetha, G. (2015). Occurrence of leaf and fruit scarring beetle, *nodostoma subcostatum* (jacoby) on banana in pest management in horticultural ecosystems, 21: 221-224.

Singh, P.P., Singh, S.P. and Mondal, S.S. (1997). Incidence and seasonal variation of banana scarring beetle, *Nodostoma subcostatum* Jacoby on banana in North Bihar.*Natl. Seminar on Orchard Management for Sustainable Production of Tropical Fruits* held on March 10-11, 1997 at R.A.U. Bihar, Pusa (India) pp. 77.

6

Biointensive Integrated Pest Management of Pomegranate

Sandeep Kumar, Priyanka Kumawat and Ponnusamy N.

Pomegranate (*Punica granatum* L.) also known *Anar* in Hindi it is belonging to the Punicaceae family, it is grown in tropical and sub-tropical regions. Originated in Indo-Burma region Iran and is extensively cultivated in the Mediterranean region since ages (Sheikh and Manjula, 2009). Pomegranate is used as table purpose, making juice, jam, jelly and paste. The fruit peel, tree stem and leaves and root bark are good source of secondary metabolites such as tannins, dyes and alkaloids (Mirdehghan and Rahemi, 2007). The edible part of the fruit is the seeds having a fleshy covering and called arils.

India is major pomegranate producing country in the world. In India main pomegranate producing states are Maharashtra, Karnataka, Gujrat, Andhra Pradesh, Madhya Pradesh, Tamil Nadu and Rajasthan. Maharashtra is leading state of pomegranate production in India. The prominent pockets where pomegranate cultivated area are concentrated are Solapur, Nashik, Sangli, Satara and ahemednagar districts of Western region of Maharashtra.

Indian total pomegranate crop covers 0.24 Mha and production is 2.87 MT in India (Anonymous 2017). The export of pomegranate from India is only 0.04 MT it is very low. Pomegranate is majorly exported to UAE (43%), Bangladesh (16%) and European Countries (14%) A few tons also go to Saudi Arabia, Russia, Thailand, Nepal, Kuwait etc.

Pomegranate production is also connected with many problems like other crops. Inherent constraints are long dry spells, less availability of resistant variety, non-availability of suitable varieties, environmental vagaries, physiological disorders, nutritional deficiencies, post-harvest losses and improper storage. Biotic constraints are like insect or pest, disease and disorder problems. In India, pomegranate is attacked by more than 45 insects (Butani, 1979). Among these constraints, losses due to pests and diseases are very high. Although, 25 to 30 per cent of total cost of production is being spent on plant protection especially pesticides, the biotic constraints could not be managed effectively.

For the successful management of any disease under normal conditions, clean sanitation, eradication of primary source, chemical biochemical and predators protection at initial stages are some of the measures recommended. However, these measures are not enough, whenever the outbreak of disease and pests occurred. Hence, thorough understanding of the disease and pests epidemiology and concrete package is necessary to address the menace effectively, so as to save the crop at large.

Among the various factors which contribute towards the growth, yield and quality of pomegranate integrated disease and pest management and Biointensive Integrated Pest Management is the most important practices and it has direct effect on production and quality. These practices in pomegranate cultivation to improve the growth yield and quality and minimizing of crops loss due to disease and pests.

Biointensive Integrated Pest Management of Pomegranate

Biointensive Integrated Pest Management incorporates ecological and economic factors into agricultural system design and decision making and addresses public concerns about environmental quality and food safety. Biointensive Integrated Pest Management is a difficult way of motto that pest prevention is created through natural and non-chemical means. Besides using pesticides, methods such as crop rotation can prevent a pest problem in our horticulture crops. So many other options include the introduction of native predators of the pests, the introduction of sterile males, or the reintroduction of natural, disease-fighting microbes into the plants or soil. Chemicals are only used as very last resorts. Sustainable agriculture and horticulture has many benefits to humans envirment as well as animals. Industrial farms create air pollution; their animals build up antibody resistance and often transmit diseases to humans. The present deals with the most recent biointensive integrated pest approaches using components such as predators, parasitoids and pathogens botanicals like biofumigation, oil cakes, compost, crop residues, FYM, green manuring and other organic amendments, physical methods (hot water treatment of planting material, soil solarization), cultural methods (crop rotation, pest resistant variety, summer ploughing, fallowing, intercropping, pruning, mulching, spacing, planting date, trap cropping, etc.) and biorational chemicals.

Integrated Pest Management

IPM plan envisages using cultural, mechanical, chemical and biological methods at the same time in combating pest's problem. At present much importance is being given on integrated approaches primarily due to forcefulness of pest population and increase demand of health safe produce. As per expected, about 20.0% of total loss to crops is attribute to insect or pests alone. Proper management of pests we take good and healthy crops.

The Benefits of Biointensive Integrated Pest Management

- Biointensive Integrated Pest Management can include reduced chemical input costs
- Reduced on-farm and off-farm environmental impacts
- More effective and sustainable pest management
- Biointensive an ecology-based IPM has the potential of declining inputs of fuel, equipment, machinery and synthetic chemicals-all of which are power intensive and more and more costly in terms of economic and environmental impact. Such of chemical and inputs reductions will benefit for the grower, society and envirment.
- The information on biointensive integrated pest management (insect, mite and nematode pests, and diseases caused by bacteria, fungi, virus/ mycoplasma) in horticultural ecosystems is very much scattered.

Insect Pests of Pomegranate

1. Anar Butterfly

Scientific name	:	*Virachola isocrates* Fabricius/ *Deudorix Isocrates*
Family	:	Lycaenidae
Order	:	Lepidoptera
Common name	:	Anar Butterfly

Biology

- **Eggs:** Laid singly on tender leaves, stalks, and flower buds.
- **Larvae:** Bark brown colour, short and stout, enclosed with short hairs, larval age lasts for 18-47 days.
- **Pupation:** Occurs either inside the injured fruits or on the stalk holding it. Pupa period lasts for 7-34 days. Total living cycle is completed in 1 to 2 months.
- **Adult:** Bluish brown butterfly and female is V shaped patch on forewing

Damaging symptoms

Pomegranate fruit borer or pomegranate butterfly is a most serious pest found all over India and common in Asia. It is most widespread, polyphagous and destructive pest. Infestation start from flowering to button stage during the month of august in monsoon, while is more during November or December in winter crop. The caterpillar bore inside the developing fruits and they feeds on internal contents of pulp and seeds. Such infested fruits are also invaded by bacteria and fungi and causing fruit to rot. Such affected fruits drop down. It is mostly prevalent during the 'mrig' *bahar* season (June to February).

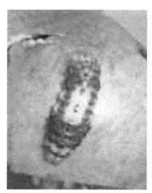   

Anar butterfly   Infection of anar butterfly on fruit

Management practices

*   Clean cultivation as weed plants serve as alternate hosts and maintenance of health and vigorous of the tree should be followed.

*   Avoidance of rainy season crop

*   Removal eggs from calyx and destruction of all the affected fruits.

*   Detect early infestation by periodically looking for drying branches.

*   Clip off calyx cup immediately after pollination followed by two applications of neem oil @ 3 %.

*   Before maturity, polythene bag young fruits cover with of 300 gauge thickness or butter paper.

*   At flowering stage, spray neem formulations 2 ml/1.

*   Spray Deltamethrin 2.8 EC (1.5 ml/litre of water) at fortnightly interval from the stage of flowering to fruit development.

- Spray Malathion 50 EC 0.1% or Azadirachtin 1500ppm at 15 days intervals commencing from initiation of flowering up to the harvesting subjected to the presence of fruit borer.
- Release *Trichogramma chilonis* @ one lakh/acre.

2. Stem Borer

Scientific Name	:	*Coelosterna spinator* (Febricius)
Family	:	Coleoptera
Order	:	Cerambycidae
Common name	:	Stem borer

Biology

- **Egg-** Egg period is 12 to 15 days,
- **Grub-** Grub 9 to 10 months
- **Pupa-** Pupal period is 16 to18 days. There is only one generation per year and longevity of beetles is 45 to 60 days.
- **Adult-** Beetles have pale yellowish-brown body with light grey elytra and are 30 to 35 mm long.

Damaging symptoms

It is a polyphagous pest of minor importance boring the stems and trunk of pomegranate trees. The caterpillar of this pest makes hole and bores into the cambium then girdle the main stem or branch causing death of the tree. It comes out at night and feeds bark. It prefers breeding in dead wood but also attacks the living branches. Generally, excreta and dry powdered material is seen near the base of plants.

Stem borer   Stem borer on stem

Management practices

- Avoid water logging and rake the soil.
- Infested trees should be uprooted and brunt, especially the root zone.
- Prune the affected fruits and buds of the trees and destroy.
- Clean the hole by removing insect's excreta with the help of iron stick. Plug the hole and bores with cotton plug dipped in petrol, chloroform or kerosene oil followed by sealing it with mud or painting with coal tar.
- Treat the holes by injecting with fenvalerate 5 ml/L or dichlorvos 10 ml /L and seal holes with clay. Spray quinalphos (0.05%) or chlorpyrifos (0.05%).

3. Bark-eating Caterpillar (*Indarbela tetraonis*)

Scientific Name	:	*Indarbela quadrinotata* walker
Family	:	Metarbedelidae
Order	:	Lepidoptera
Common name	:	Bark-eating Caterpillar

Biology

- **Egg:** They are oval, reddish in colour
- **larva:** caterpillars are pinkish white with brown spots and are about 40 mm long
- **Pupa:** Pupae are chestnut-brown in colour and 22 to 28 mm long.
- **Adult:** Moths are white with pairs of small black dots on thorax, numerous small black spots and streaks on fore wings and few black spots on posterior edges of hind wings.

Bark eating caterpillar eggs are laid in clusters of 18-25 under wobbly bark or in cracks and crevices from April to June. They come up with in 8-11 days with the larval duration of 9-10 months. Pupal period extends to 3 to 4 weeks. Total life cycle completed 4-5 months in south India and more than a year in north India. It completes one generation per year.

Damaging symptoms

This pest is a polyphagous pest. The pest has been reported to infest 70 plant species across fruits, forests and avenue plantations (Verma and Khurana, 1978). *Indarbela tetranis* and *Indarbela quadrinotata* have been recorded boring the bark of pomegranate tree and feeding inside. *Indarbela tetranis* is comparatively more common and more harmful to pomegranate trees. The caterpillar is night-time in nature and feeds on the bark of the host plant under a frass ribbon and remains hidden inside the galleries formed at the forking points on trunk and branches during the day. Tree become weak and does not bear fruits. Older trees and not well maintained orchard are more prone to this pest. Peak activity period is September-October. Moths are white with pairs of small black dots on thorax, numerous small black spots and streaks on fore wings and few black spots on posterior edges of hind wings.

Bark-eating Caterpillar   Make frass ribbon by insect

Management practices

- Maintain clean orchards by avoiding overcrowding of trees

- Clean the webs around the affected portion

- Inject kerosene oil chloroform and petrol into the holes and seal with mud (Srivastava, 1972).

- Inject larval holes with quinalphos @ 0.01% or fenvalerate @ 0.05%. Spray with carbaryl @ 0.04% or dichlorovos @ 0.08% on the stem or on affected part.

Sap Sucking Insects

a) Thrips (*Scirtothrips dorsalis* and *Rhipiphorothrips cruentatus* Hood)

Scientific Name   :   *Rhipiphorothrips cruentatus, Scirtothrips dorsalis*

Family   :   Thripidae

Order   :   Thysanoptera

Biology

- **Egg:** Female lays on an average 50 dirty white bean-shaped eggs on the under surface of leaves.

- **Nymph:** The incubation period is 3-8 days. Newly hatched nymphs are reddish and turn yellowish brown as they grow.

- **Pupa:** Pupa period lasts 2-5 days.

- **Adult:** The adult sare minute, slender, soft bodied insects with heavily fringed wings, blackish brown with yellowish wings and measure 1.4 mm long. Adult is straw yellow in colour.

Thrips

Thrips scrapping on fruits

Damaging symptoms

Nymphs and adults of all the species were seen on the under surface of the leaves, on fruits and flowers. The lacerating and sucking by the thrips resulted in shriveling of leaves and fruits. Scarring of rind was also observed on fruits due to desapping, resulting in decreased marketability of fruits (Ananda *et al.*, 2007). The incidence of this pest is mainly seen from July to October with the peak period in September.

Damaging fruits

b) Mealybug

Scientific Name	:	*Planococcus lilacinus* Cockrell
Family	:	Pseudococcidae
Order	:	Hemiptera

Biology

- **Egg:** The female lays eggs in groups which lay beneath the body in a loose ovisac of waxy fibers. Fecundity (egg number) ranged from 109 to 185 per female. The oviposition period is 20-29 days. The incubation period is 3-4 hours.

- **Nymph:** Female and male nymphs molt 3 and 4 times, respectively, and the development period varies from 26-47 and 31- 57 days, respectively.

- **Adult:** Longevity of the adult female is 36-53 days and for the male is, 1-3 days.

Mealy bug   Mealy bug infection on fruit

Damaging symptoms

Mealy bugs saw attacking fruits and their stalk on pomegranate plant. Nymph and adult both stage damage to pomegranate by sucking the sap from the leaves, flowers and fruits which resulted in yellowing of leaves and shedding of flowers and tender fruits (Ananda *et al.*, 2007). The market value of fruit is reduced and unfit for market.

c) Aphid

Scientific Name	:	*Aphis punicae* Passerini
Family	:	Aphididae
Order	:	Hemiptera

Biology

- **Egg:** Egg period is one or two days. Young aphids are called nymphs.

- **Nymph:** Nymphs are oval or slightly elongated, reddish brown with six segmented antennae.

- **Adult:** Small yellowish-green typically colonizing the upper sides of mature leaves of pomegranate, concentrated along the midribs and around the leaf margins also found on flowers but rarely on fruits.

Aphid   Aphid infection on flower

Damaging symptoms

Both nymphs and adults of which sucked sap from the tender shoots, leaves, flowers and fruits of pomegranate plant.The infestation resulted in yellowing of leaves and sticky to touch. This specific pest of pomegranate was also recorded by Butani (1976) in India and Balikai (2000) from northern Karnataka.

d) Whitefly

Scientific Name : White fly, *Siphoninus phillyreae* Haliday

Spiralling white fly, *Aleurodicus disperses* Russell

Family : Aleyrodidae

Order : Homoptera

Biology

- **Egg**: Eggs are laid in a circle on the underside of leaves.

- **Nymph**: Short glass like rods of wax along the sides of the body

- **Adult**: Powdery white, active during early morning hours.

Damaging symptoms

The ash whitefly is lesser in size compared to spiralling whitefly. The female of spiralling whitefly laid eggs in a spiral manner on lowers surface of leaves. Both adults and nymph sucked the sap from leaves and caused yellowing and fall down. Sooty mould development was also seen on the honeydew excreted by whiteflies on the upper surface of leaves. Severe incidence of ash whitefly was also recorded first time on pomegranate by Nair (1978) and Balikai *et al.* (1999) in northern Karnataka which supports the present findings.

e) Scale insects

Scientific Name : *Parasaissetia nigra* Neitn
Family : Coccidae
Order : Hemiptera

Damaging symptoms

In case of heavy infestation, black sooty mould formation was noticed on leaves. These two species of scale insects were found to infest pomegranate stem and foliages. The scale insect *P. nigra* was reported by Karuppuchamy (1994) in Tamil Nadu and Jadav and Ajri (1985) in Maharashtra on pomegranate. But this pest was a new record noticed on pomegranate from Karnataka.

Sap sucking insects management practices

* Keep basin clean.
* Maintain adequate aeration by proper training and pruning
* Prun the affected parts of the plant and destroy.
* Detect early infestation by periodical monitoring for drying branches.
* Use yellow/blue sticky traps @ 4-5 traps/acre
* For mealy bugs spray Durmet/Durban 20 EC (chloropyriphos) @ 3.75 ml in one liter of water on appearance of mealy bugs and scale insects.
* Spray of 500ml Rogor 30 EC (dimethoate) in 500 liter of water for control of white flies, aphid and thrips.
* Biological control for mealy bugs released *Cryptolaemus montrouzieri* near the site of mealybug @ 10/tree natural enemies of mealy bug.
* Biological control for aphid release first instar larva of *Chrysoperla zastrowi sillemi* @ 15 / flowering branch (four times) at 10 days interval from flower initiation during April.

References

Anonymous, (2017). *Indian Horticulture Database* 2016 published from National Horticultural Board. 15-469.

Ananda, N., Kotikal, Y.K., and Balikai, R.A. (2007). Sucking insect and mite pests of pomegranate and their natural enemies.

Balikai, R.A. (2000). Status of pomegranate pests in Karnataka. Pest Manag. *Hortic. Ecosystem*, 6: 65-66.

Balikai, R.A., Biradar, A.P. and Teggelli, R.G. (1999). Severe incidence of pomegranate whitefly, *Siphoninus phillyreae* (Haliday) in northern Karnataka. *Insect Environ.* 5: 76.

Butani, D. K. (1976). Insect pests of fruit crops and their control on pomegranate. *Pestology*, 10: 23-26.

Butani D. K. (1979). Pests of pomegranate. In: Insects and fruits. *Periodical Expert Book Agency,* Delhi, 228.

Gurr, G. M, Wratten, S. D and Altieri MA (2004a) Ecological Engineering for Pest Management Advances in Habitat Manipulation for Arthropods. CSIRO Publishing, Collingwood, Australia.

Gurr G.M., Wratten S.D. and Altieri M.A. (2004b). Ecological Engineering: a new direction for pest management, *AFBM Journal* 1: 28-35.

http://agropedia.iitk.ac.in/content/

Jadhav, S.S. and Ajri, D.S. (1985). Chemical control of scale insects on pomegranate in Maharashtra. *Indian J. Ent.* 4: 63-65.

Karuppuchamy, P. (1994). Studies on the management of pests of pomegranate with special reference to fruit borer, Virachola isocrates (Fab.). Ph.D. Thesis, Tamil Nadu Agril. Univ., Coimbatore (India).

Mirdehghan, S.H. and Rahemi, M. (2007). Seasonal changes of mineral nutrient and phenolic in pomegranate (*Punica granatum* L.) fruit. *Sci. Hortic.* 111, 120-127.

Nair, M. R.G.K. (1978). Insects and mites of crops in India, Indian Council Agril. Res., New Delhi, 40.

NCIDM (2013). pests of fruits (Banana, Mango and Pomegranate): 'E' Pest Surveillance and Pest Management Advisory 36-39.

Sheikh, M.K., Manjula, N. (2009). Effect of split application of N and K on growth and fruiting in 'Ganesh' pomegranate (*Punica granatum* L.). *Acta Hort.* 818, 213-217.

Srivastava, O.S. (1972). Chemical control of bark eating caterpillar, Indarbela quadrinotata Walker (Lepidoptera: Metarbelidae) in guava trees. *Indian Journal of Agricultural Sciences*, 42(9): 847-848.

Verma, A.N. and Khurana, A.D. (1978). Bark eating borers (Indarbela spp.). The pest of fruit, avenue and forest trees in India and South East Asia. *Haryana Journal of Horticultural Science*, 7(1-2): 40-46.

7

Biointensive Integrated Pest Management of Papaya

Nagendra Kumar, Sandeep Kumar and Priyanka Kumawat

Introduction

Papaya is a tropical tree fruit crop that cannot tolerate freezing temperatures. It has commercial importance because of its high nutritive and medicinal value. Papaya cultivation had its origin in South Mexico and Costa Rica. Total annual world production is estimated at 6 million tonnes of fruits. India leads the world in papaya production with an annual output of about 3 million tonnes. Other leading producers are Brazil, Mexico, Nigeria, Indonesia, China, Peru, Thailand and Philippines. Papaya fruit are borne on peduncles attached to the main trunk of the papaya tree and are clustered, starting a few feet above the ground and extending up to a foot below the top of the plant adequate water is essential to the uninterrupted growth and production of papaya fruit. There is number of factors which affect quality and quantity of papaya production. Out of insect pest is one of the important factors to reduction of papaya fruit production. List of insect pests of papaya given as below:

Common Name	Scientific Name	Order	Family
Papaya Mealy bug	*Paracoccus marginatus*	Hemiptera	Pseudococcidae
Grasshopper	*Poecilocerus pictus*	Orthoptera	Pyrgomorphidae
Whitefly	*Bemisia tabaci*	Hemiptera	Aleyrodidae
Scale insect	*Aspidiotus destructor*	Hemiptera	Diaspididae
Aphid	*Myzus persicae*	Hemiptera	Aphididae
Fruit fly	*Bactrocera cucurbitae*	Diptera	Tephritidae

1. Papaya Mealybug

Damage symptoms

- Papaya mealybug infestation appears on above ground parts on leaves, stem and fruits as clusters of cotton-like masses.

- The insect sucks the sap by inserting its stylets into the epidermis of the leaf, fruit and stem. While feeding, it injects a toxic substance into the leaves, resulting in chlorosis, plant stunting, leaf deformation or crinkling, early leaf and fruit drop, and death of plants.

- The honeydew excreted by the bug results in the formation of black sooty mould which interferes in the photosynthesis process and causes further damage to the crops.

- Heavy infestations are capable of rendering fruit inedible due to the buildup of thick white waxy coating.

Biology

- Papaya mealybugs are most active in warm, dry weather. Females usually lay 100 to 600 eggs.

- Eggs are greenish yellow and are laid in an ovisac sac that is three to four times the body length and entirely covered with white wax.

- Egg-laying usually continuous over a period of one to two weeks.

- Eggs hatch in about 10 days, and nymphs or crawlers begin to actively search for feeding sites.

- Adult males may be distinguished from other related species by the presence of stout fleshy setae on the antennae and the absence of fleshy setae on the legs.

- Females have three instars whereas males have four instars. Males have longer development time (27-30 days) than females (24-26 days).

Factors responsible for high population buildup

- Wax layer and waxy fibres over the ovisac and body of mealybug nymphs and adult females protect them from adverse environmental conditions and routine chemical pesticides.
- Availability of alternate hosts / weeds around fields not cared by cultivators. Movement of crawlers through air, irrigation water or farm equipment helps in fast spread of the mealybug from infested field to healthy fields.

BIP Management Strategies

Cultural and Mechanical

- Monitoring and scouting to detect early presence of the mealybug
- Pruning of infested branches and burning them
- Removal and burning of crop residues
- Removal of weeds/alternate host plants like Hibiscus, Parthenium etc. in and nearby crop
- Avoiding the movement of planting material from infested areas to other areas
- Avoiding flood irrigation
- Prevention of the movement of ants and destruction of already existing ant colonies
- Sanitization of farm equipment before moving it to the uninfested crop
- Application of sticky bands or alkathene sheet or a band of insecticide on arms or on main stem to prevent movement of crawlers.
- Preventive: In tapioca, stems are stocked for propagative purpose in the farms.
- These planting materials often carry mealybug infestation, if the previous year's crop was already infested.

Biological control

- Natural enemies of the papaya mealybug include the commercially available mealybug destroyer *Cryptolaemus montrouzieri*, ladybird beetles, lacewings, hover flies, Scymnus sp. and certain hymenopteran and dipteran parasitoids.

Conservation of these natural enemies in nature plays important role in reducing the mealybug population.

- There is a need to conserve the native predators of the pest. Australian ladybird beetle (*C. montrouzieri*) predates on mealybugs, eating 3,000-5,000 mealybugs in various life stages and is released @ 10 beetles per tree or @ 5000 beetles/ ha.

Chemical control

- Regular monitoring of the crop for mealybug infestation and its natural enemies.

- Spot application of insecticide immediately after noticing mealybug on some plants in the crop field.

- If the activities of natural enemies are not observed, use of botanical insecticides such as neem oil (1 to 2%), NSKE (5%), or Fish Oil Rosin Soap (25g/litre of water) should be the first choice.

- Avoid repeating the use of the same chemical insecticide as there are chances for development of resistance in the pest.

- Drenching soil with chlorpyriphos around the collar region of the plant to prevent movement of crawlers of mealybug and ant activity is useful.

2. Grasshopper

Damage symptom

- It is primarily a defoliator of ak plant (*Calotrophis gigantee*), but also causes damage to papaya, ber, banana, guava, melon, peach, citrus and other plants and vegetables.

- Both adult and nymphs are voracious eaters. They defoliate the plant.

- In severe infestation they even eat the bark of the papaya tree.

- The infestation starts in April and continues till October. Maximum damage is caused during July-August.

Life History

- Male and female mates after emergence, Mating lasts for 5-7 hours.

- Female starts laying egg three to four weeks after mating, for egg lying female thrust its abdomen deep into the soil and deposit 145 to 170 eggs, about 18-20 cm beneath the surface.

- The eggs are orange coloured, elongated and are laid in spiral fashion.

- Female covers the egg mass with frothy secretion.
- During autumn the nymph hatches out from the eggs in about a month and become adult in 55 to 60 days
- The eggs lie in September-November hatches into nymph after about four months and become adult after a nymphal period of about 2 1/2 months.

BIP Management Strategies

- Deep ploughing of fields during summer to control nematodes population and weeds.
- Soil solarization
- Timely sowing should be done.
- Field sanitation.
- Destroy alternate host plants
- Plant tall border crops like maize, sorghum or millet to reduce whitefly and aphid population.
- Crop rotation with non-host crops
- Removal of weeds and alternate host plants such as *Hibiscus,* bhindi, custard apple, guava etc.

3. Whitefly

Damage Symptoms

- Both the adults and nymphs suck the plant sap and reduce the vigor of the plant.
- In severe infestations, the leaves turn yellow and drop off.
- When the populations are high they secrete large quantities of honeydew, which favors the growth of sooty mould on leaf surfaces and reduces the photosynthetic efficiency of the plants.

Biology

- The females mostly lay eggs near the veins on the underside of leaves. Each female can lay about 300 eggs in its lifetime.
- Upon hatching, the first instar larva (nymph) moves on the leaf surface to locate a suitable feeding site. Hence, it is commonly known as a "crawler." If then inserts its piercing and sucking mouthpart and begins sucking the plant sap from the phloem.

- Adults emerge from puparium through a T-shaped slit, leaving behind empty pupal cases or exuviae.

- The whitefly adult is a soft-bodied, and moth-like fly. The wings are covered with powdery wax and the body is light yellow in color.

- The wings are held over the body like a tent. The adult males are slightly smaller in size than the females. Adults live from one to three weeks.

BIP Management Strategies

- Water sprays may also be useful in dislodging adults.

- A small, hand-held, battery-operated vacuum cleaner has also been recommended for vacuuming adults off leaves. Vacuum in the early morning or other times when it is cool and whiteflies are sluggish. Kill insects by placing the vacuum bag in a plastic bag and freezing it overnight. Contents may be disposed of the next day. Fumigating with a small petrol soacked cotton ball.

- Plant tall border crops like maize, sorghum or millet to reduce whitefly and aphid population. Use yellow sticky traps for aphids and whitefly @ 4-5 traps/acre.

- Apply 5% neem seed kernel extract (NSKE) or groundnut oil @ 1-2% on to the plants to manage.

4. Scale Insect

Biology

- Freshly laid eggs are smooth, elongate and whitish, becoming pale yellow over time.

- Eggs are laid under the scale cover around the body of the female.

    Newly hatched nymphs (also called crawlers) are free-moving and recognized by the presence of legs, antennae, and a pair of bristles at the tip of abdomen. Crawlers are light green to yellowish brown, translucent and somewhat oblong,

- Females remain pale yellow, circular and somewhat transparent in the second instar, which lasts for 8-10 days.

- Development of male characteristics starts in the middle of the second instar. Male covers become reddish brown and more elliptical in shape, and then transform in stages into the pre-pupal, pupal and adult stages. The second instar lasts for 5-8 days for males. The male pre-pupal and pupal stages are spent under the scale produced by the second instar.

- Adult female coconut scales have a circular or broadly oval cover. The cover is flat and translucent with a subcentral pale exuviae.

- Adult male coconut scales are small, two-winged, reddish, gnat-like insects with eyes, antennae, three pairs of legs and long appendages. Adult males do not feed and are short lived.

BIP Management Strategies

- Pruning and training of fruit trees and proper disposal of infested leaves, branches and twigs will help control scale insects on nursery plants and trees.

- Burn the remaining debris.

- Removal of attendant ants may permit natural enemies to control the insect.

- Excessive use of plant fertilizers contributes to scale outbreaks.

- *Aphytis melinus* and *Aphytis lingnanensis* (Hymenoptera: Aphelinidae) are the most common parasitoid species controlling the coconut scale.

- *Aphytis melinus* and/or another parasitoid, *Comperiella bifasciata* (Hymenoptera: Encyrtidae), were introduced into Hawaii, California, Argentina and other regions for the control of coconut scale.

- Coccinellid ladybird beetle predators of coconut scale include *Chilocorus* spp., *Telsimia nitida* Chapin, *Pseudoscymnus anomalus* Chapin and *Cryptognatha* spp.

5. Aphid

Damage Symptoms

- Aphids are one of the minor pests of papaya causing damage to the plant by way of sucking the juice with their piercing-sucking mouth parts.

- They feed on the undersurface of leaves and cause them to become curled and crinkled.

- The petioles of heavily infested leaves generally droop downward.

PRSV symptom

- They also secrete honeydew which attracts ants and thereby promote the growth of sooty mould on the papaya plant.

- Aphids feed by sucking sap from their hosts. Extensive feeding causes distortion of young leaves and shoots and premature dropping of fruit.

- Besides causing direct damage to the papaya plants by feeding on them, they also act as vector of papaya ring spot.

Biology

- Aphids are soft-bodied insects of 20 to 30 mm long having long antennae and legs.
- They may be green, yellow or black in colour with and without wings.
- An unfertilized female aphid can produce 60 to 100 nymphs in a life span of about a month.
- The nymphs mature and become adult in about a week.
- There are many generations of this aphid throughout the year.
- Longevity may be affected by temperature, type of egg laying or live births, and plant host. Studies in cooler temperatures report the life cycle lasting up to 50 days (Toba, 1964).
- Populations are larger during periods of adequate rainfall and smallest during hot, dry weather.

BIP Management Strategies

Cultural Control

- Remove crop residues and weed hosts prior to planting new crops.

Biological Control

- Most effective natural enemies are the predatory syphid maggots, *Allograpta sp.,* lady beetles and parasitic wasps.
- Release larvae of green lacewing bug (*Chrysoperla zastrowi sillemi*) @ 4,000 larvae/acre.

Chemical Control

- Apply 5% neem seed kernel extract (NSKE) or groundnut oil @ 1-2% on to the plants to manage the vector population.

6. Fruit fly

Damage Symptom

- They oviposite in fruit and soft tissues of vegetative parts of hosts and feeding by the larvae and larval feeding damage in fruits is the most damaging.
- They decompose plant tissue by invading secondary microorganisms.

- Mature attacked fruits develop a water soaked appearance. Young fruits become distorted and usually drop.

- The larval tunnels provide entry points for bacteria and fungi that cause the fruit to rot.

- These maggots also attack young seedlings, succulent tap roots of watermelon, and stems and buds of host plants such as cucumber, squash and others.

Biology

- The eggs of the melon fly are slender, white and measure 1/12 inch in length. Eggs are inserted into fruit in bunches of 1 to 37. They hatch in 2 to 4 days.

- There are 3 larval stages for this insect. The larvae, or maggots, are typical fruit fly shape: cylindrical-maggot, elongate, narrowed and somewhat curved downward at the end and mouth hooks at the head. These maggots reach approximately 1/2 inch in length upon maturity. Refer to Heppner (1989) for a detailed description of larvae. The larval period lasts from 6 to 11 days, with each stage lasting 2 or more days. Duration of larval development is strongly affected by host.

- Pupae occur in the soil beneath the host plant. They are 1/5 to 1/4 inch long, elliptical and dull white to yellowish brown in color. They are distinctly ringed by narrow yellow bands around each segment. During warm weather the pupal stage lasts 9 to 11 days.

- Adult melon flies are slightly larger than houseflies. They measure 1/3 to 1/2 inch long with a wingspan of 1/2 to 3/5 inch.

- Oviposition occurs about 10 days after emergence and continues at intervals. One female may deposit up to 1,000 eggs, although 300 eggs total are estimated in natural conditions.

- Females prefer to oviposit in new plant growth such as young seedlings, growing tips, and developing ovaries of all cucurbits except young cucumbers. Ripe fruits are preferred; green fruits are sometimes used. Because of their high egg laying capacity and mobility, each female is capable of destroying large numbers of fruit in her lifespan. Adults generally live for 10 months to a year.

BIP Management Strategies

Mechanical and Physical Control

- Wrapping developing fruit with a protective covering and the use of baited traps. Baited traps are used to destroy adults. Shrubs within 100 yards of larval hosts may be used advantageously in placing traps.

- Hot water treatment of fruit at 48 ± 1 °C for 45 min.

- Set up fly trap using methyl eugenol. Prepare methyl eugenol 1ml/L of water + 1 ml of malathion solution. Take 10 ml of this mixture per trap and keep them at.

Cultural Control

- Collect and dispose off infested and fallen fruits after harvest to prevent further multiplication and carry-over of population.

- Ploughing of papaya plantation during November-December to expose pupae to sun's heat which kills them.

- Use of trap crops to reduce fruit fly.

Biological Control

- *Opius longicaudatus* var. *malaiaensis* (Fullaway), *O. vandenboschi* (Fullaway) and *O. oophilus* (Fullaway),parasites are primarily effective on fruit flies.

References

Agarwal, M. L., D. D. Sharma and O. Rahman (1987). Melon Fruit-Fly and Its Control. *Indian Horticulture*. 32(3): 10-11.

Ahmad R, Ghani MA. (1972). Studies on *Aspidiotus destructor* Sign. (Hemiptera: Diaspididae) and its parasites, *Aphytus melinus* Debach (Hymenoptera: *Aphelinidae*) and *Pakencyyrtus pakistanensis* Ahmad (Hymenoptera: *Encyrtidae*) in Pakistan. *Commonwealth Institute of Biological Control Technical Bulletin* 15: 51-57.

Anonymous. (1966). Green Peach Aphid, *Myzus persicae* 4th Edition. New South Wales Department of Agriculture Entomology Branch, Insect Pest Bulletin No. 16. Sydney: V.C.N. Blight, Government Printer. 6 pages.

Ben-Dov Y, Miller DR, Gibson GAP. (2014). ScaleNet Aspidiotus destructor Signoret. (12 February 2015).

Bennett FD, Alam MM. (1985). An annotated check-list of the insects and allied terrestrial arthropods of Barbados. Bridgetown, Barbados; Caribbean Agricultural Research and Development Institute, pp. 6-81.

Bess, H. A., R. van den Bosch and F. H. Haramoto. (1961). Fruit Fly Parasites and Their Activities in Hawaii. Proc. *Hawaiian Entomol. Soc.* 27(3): 367-378.

Blackman, R. L. and V. F. Eastop. (1984). *Myzus persicae* (Sulzer). In: Aphids on the Worlds Crops: An Identification and Information Guide. John Wiley and Sons: Chichester, New York, Brisbane, Toronto, Singapore. 466 pages.

Burbutis, P.P., C.P. Davis, L.P. Kelsey and C.E. Martin. (1972). Control of Green Peach Aphid on Sweet Peppers in Delaware. *J. Econ. Entomol.* 65(5): 1436-1438.

Chua TH, Wood BJ. (1990). Other tropical fruit trees and shrubs. In: D. Rosen (ed.), armored scale insects, their biology, natural enemies and control. World Crop Pests. Elsevier, Amsterdam, the Netherlands, 4: pp 543-552.

Claps LE, Wolff VRS, González RH. (2001). Catálogo de las Diaspididae (Hemiptera: Coccoidea) exóticas de la Argentina, Brasil y Chile. Revista de la Socieded *Entomológica Argentina* 60: 9-34.

Danzig EM, Pellizzari G. (1998). Diaspididae. In: Kozár F, (Ed.), Catalogue of Palearctic Coccoidea. Hungarian Academy of Sciences. Budapest, Hungary: pp. 172-370.

Davidson JA, Miller DR. (1990). Ornamental plants. In Rosen, D (ed.) Armored scale insects, their biology, natural enemies and control. Elsevier, Amsterdam, the Netherlands, 4 (B): pp. 603-632.

DeBach P. (1974). Coconut scale in Fiji. In Biological Control by Natural Enemies. DeBach P (ed.). Cambridge Univ. Press, Cambridge UK, pp. 133-135.

EENY-302, one of a series of Featured Creatures from the Entomology and Nematology Department, Florida Cooperative Extension Service, Institute of Food and Agricultural Sciences, University of Florida, Published: August 2003.

Elmer, H.S. and O.L. Brawner. (1975). Control of Brown Soft Scale in Central Valley. *Citrograph.* 60(11): 402-403.

Flint, M. L. (1985). Green Peach Aphid, *Myzus persicae*. pp. 36-42. Integrated Pest management for Cole Crops and Lettuce. University of California Publication 3307. 112 pages.

Follet PA. (2006). Irradiation as a phytosanitary treatment for *Aspidiotus destructor* (Homoptera: Diaspididae). *Journal of Economic Entomology* 99: 1138-1142.

Ghauri M. (1962). The morphology and taxonomy of male scale insects (Hemiptera: Coccoidea). British Museum (Natural History). Adlard and Son, Dorking, UK. 221 pp.

Heppner, J. B. (1989). Larvae of Fruit Flies. V. *Dacus cucurbitae* (Melon Fly) (Diptera: Tephritidae). Fla. Dept. Agric. & Consumer Services, Division of Plant Industry. Entomology Circular No. 315. 2 pages.

Heu RA. (2002). Distribution and host records of agricultural pests and other organisms in Hawaii. USA: State of Hawaii Department of Agriculture.

Hill, D.S. (1983). *Dacus cucurbitae* Coq. pp. 391. In Agricultural Insect Pests of the Tropics and Their Control, 2nd Edition. Cambridge University Press. 746 pages.

Hill, D. S. (1983). *Myzus persicae* (Sulz.). pp. 202. In Agricultural Insect Pests of the Tropics and Their Control, 2nd Edition. Cambridge University Press. 746 pages.

Ishii, M. (1972). Observations on the Spread of Papaya Ringspot Virus in Hawaii. *Plant Dis. Rep.* 56(4): 331-333.

Kessing JLM, Mau RFL. (2007). Aspidiotus destructor (Signoret). Crop Knowledge Master, University of Hawaii, http://www.extento.hawaii.edu/ kbase/crop/type/a_destru. htm (12 February 2015).

Kumari DA, Anitha V, Lakshmi BKM. (2014). Evaluation of insecticides for the management of scale insect in mango (*Mangifera indica*). *International Journal of Plant Protection* 7: 64-66.

Lall, B. S. (1975). Studies on the Biology and Control of Fruit Fly, *Dacus cucurbitae* COQ. *Pesticides*. 9(10): 31-36.

Liquido, N.J., R.T. Cunningham, and H.M. Couey. (1989). Infestation Rate of Papaya by Fruit Flies (Diptera: Tephritidae) in Relation to the Degree of Fruit Ripeness. *J. Econ. Ent.* 82(10): 213-219.

Lockwood, S. (1957). Melon Fly, *Dacus cucurbitae*. Loose-Leaf Manual of Insect Control. California Department of Agriculture.

Marsden, D. A. (1979). Insect Pest Series, No. 9. Melon Fly, Oriental Fruit Fly, Mediterranean Fruit Fly. University of Hawaii, Cooperative Extension Service, College of Tropical Agriculture & Human Resources.

Metcalf, C. L., and W. P. Flint. (1962). Destructive and Useful Insects Their Habits and Control, Fourth Edition. Revised by: R. L. Metcalf. McGraw-Hill Book Company, Inc. New York, San Francisco, Toronto, London. pp. 754.

Moreno DS. (1972). Location of the site of production of the sex pheromone in the yellow scale and the California red scale. *Annals of the Entomological Society of America*. 65: 1283-1286.

Nakahara S. (1982). Checklist of the Armored Scales (Homoptera: Diapididae) of the Conterminous United States. Washington, USA: USDA, Animal and Plant Health Inspection Service, Plant Protection and Quarantine, 110 pp.

Namba, R. and S. Y. Higa. (1981). Papaya Mosaic Transmission as Affected by the Duration of the Acquisition Probe of the Green Peach Aphid - *Myzus persicae* (Sulzer). *Proc. Hawaiian Entomol. Soc.* 23(3): 431-433.

Narayanan, K., S. Jayaraj, and T.R. Subramaniam. (1975). Control of the Diamond Back-Moth, *Plutella xylostella* L. and the Green Peach Aphid, *Myzus persicae* Sulzer with Insecticides and *Bacillus thuringiensis* var. *thuringiensis* Berliner. *Madras Agric. J.* 62(8): 498-503.

Nishida, T and H. A. Bess. (1957). Studies on the Ecology and Control of the Melon Fly *Dacus* (Strumeta) *cucurbitae* Coquillett (Diptera: Tephritidae). Hawaii Agric. Exp. Station Tech. Bull. No. 34. pages 2-44.

Nishida, T. and F. Haramoto. (1953). Immunity of Dacus cucurbitae to Attack by Certain Parasites of *Dacus dorsalis. J. Econ. Ent.* 46(1): 61-64.

Papaya Mealybug and its Management: Published by Directorate of Centre for Plant Protection Studies, Tamil Nadu Agricultural University, Coimbatore 641003.

Purcifull, D. E. Hiebert and J. Edwardson. (1984). Watermelon Mosaic Virus 2. CMI/AAB Descriptions of Plant Viruses No. 293 (No. 63 revised).

Purcifull, D., J. Edwardson, E. Hiebert, D. Gonsalves. (1984). Papaya Ringspot Virus. CMI/AAB Descriptions of Plant Viruses, No. 292 (No. 84 revised).

Salahud din, Rahman HU, Khan I, Daud MK. (2015). Biology of coconut scale, *Aspidiotus destructor* Signoret (Hemiptera: Diaspididae), on mango plants (*Mangifera* sp.) under laboratory and greenhouse conditions. Pakistan Journal of Zoology, in press.

Tanwar, R.K., Jeyakumar, P. and Monga, D. (2007) Mealybugs and their management, Technical Bulletin 19, August, 2007, National Centre for Integrated Pest Management, New Delhi.

Tao C. (1999). List of Coccoidea (Homoptera) of China. Taichung, Taiwan: Taiwan Agricultural Research Institute, Wufeng, 1-176.

Taylor THC. (1935). The Campaign against *Aspidiotus destructor*, Signoret in Fiji. *Bulletin of Entomological Research* 26: 1-102.

Thangamalar, A., Subramanian, S. and Mahalingam, C. A. (2010). Bionomics of papaya mealybug, Paracoccus marginatus and its predator Spalgius epius in mulberry ecosystem, *Karnataka J. Agric. Sci.*, 23(1): 39-41.

Timberlake, P.H. (1918). Notes on Some of the Immigrant Parasitic Hymenoptera of the Hawaiian Islands. *Proc. Hawaiian Entomol. Soc.* 3(5): 399-407.

Toba, H. H. (1962). Studies on the Host Range of Watermelon Mosaic Virus in Hawaii. *Plant Dis.* 46: 409-410.

Toba, H. H. (1963). Vector-Virus Relationships of Watermelon Mosaic Virus and the Green Peach Aphid, *Myzus persicae. J. Econ. Ent.* 56: 200-205.

Toba, H. H. (1964). Life-History Studies of *Myzus persicae* in Hawaii. *J. Econ. Entomol.* 57(2): 290-291.

UK CAB International, (1966). *Aspidiotus destructor.* Distribution maps of plant pests, June. Wallingford, UK: CAB International, Map 218. (10 February 2015).

van Emden, H. F, V. F. Eastop, R. D. Hughes and M. J. Way. (1969). The Ecology of *Myzus persicae. Ann. Rev. Ento.* 14: 197-270.

Vargas, R.I. and J.R. Carey. (1990). Comparative Survival and Demographic Statistics for Wild Oriental Fruit Fly, Mediterranean Fruit Fly, and Melon Fly (Diptera: Tephritidae) on Papaya. *J. Econ. Ent.* 83(4): 1344-1349.

Waterhouse D.F., Norris K.R. (1987). *Aspidiotus destructor* Signoret in Biological Control Pacific prospects. Inkata Press, Melbourne. Chapter 8: 454 p.

Watson GW, Adalla CB, Shepard BM, Carner GR. (2015). *Aspidiotus rigidus* Reyne (Hemiptera: Diaspididae): a devastating pest of coconut in the Philippines. *Agricultural and Forest Entomology* 17: 1-8.

Williams D.J., Watson G.W. (1988). The scale insects of the tropical south pacific region. Part 1, the armored scales (Diaspididae). CAB International, Wallingford, UK. 290 p.

Zimmerman, E.C. (1948). *Myzus persicae* (Sulzer). pp. 116-118. In: Insects of Hawaii, A Manual of the Insects of the Hawaiian Islands, including Enumeration of the Species and Notes on their Origin, Distribution, Hosts, Parasites, etc. Volume 5, Homoptera: Aphididae. University of Hawaii Press, Honolulu. 464 pages.

8

Biointensive Integrated Pest Management of Guava

Beer Bahadur Singh

Introduction

The guava fruit is a good source of vitamin C, pectin, calcium and phosphorus. The fruit is used for the preparation of processed products like jams, jellies and beverages. Guava jelly puree is very popular for its attractive purplish-red colour, pleasant taste and aroma. The puree can be used in juice, cakes, puddings, sauces, ice-cream, jam and jelly. Fruits can be preserved by canning as halves or quarters, with or without seed core (shells). Good quality salad can be prepared from the shell of ripe fruits. Their cultivation in India about 1.32 lakh ha area with production is 17.12 lakh tonnes. Biointensive pest management is use to major emphasis on conservation and enhancement of natural enemies and utilization of all compatible methods for achieving effective economical and safe pest control. These methods are the most appropriate in horticulture and in protected cultivation. The major aim of Biointensive pest management is to provide guidelines and options for the effecting management of pests and beneficial organisms in an ecological context. It will help to reduce the dependence on chemical pesticides and ecological deterioration. Biointensive pest management includes bio-pesticides derived from microbes, parasitoides, predators, botanicals and all conventional non-chemical methods or use of need based. Large number of insect pests has been reported to occur on guava at various growth stages. More than 80 species of insects and mites have recorded on affecting the growth and production. About three species of fruit flies, *Bactrocera dorsalis, B. Cucurbitae* and *B. Zonata* found to attack guava fruits, *B. Dorsalis* being the dominant. Maximum activity of fruit flies is observed during August to December reaching its peak during September. Two species of bark eating caterpillars, *Indarbela quadrinota* and *I. Tetraonis* are commonly found in the region. Sucking pests included Mealy scale, *Chloropulvinaria psidii,* Mealy bugs Ferrisia virgata, *Plannococcus citri*, Tea mosquito bugs,

Helopeltis antonii, aphid, jassids etc (Azad Thakur *et al.*, 2009). The major pests of guava that are economically damage on this crop given below:

S.No.	Common Name	Scientific Name	Family	Order
1.	Fruit fly	*Bactrocera spp.*	Tephritidae	Diptera
2.	Tea mosquito bug	*Helopeltis antonii*	Miridae	Hemiptera
3.	Fruit borer	*Virachola isocrates,*	Pseudococcidae:	Hemiptera
4.	Bark caterpillar	*Indarbela tetraonis*	Metarbelidae	Lepidoptera
5.	Mealy bug	*Ferrisia virgata*	Pseudococcidae	Hemiptera
6.	Scale insect	*Chloropulvinaria psidii*	Coccidae	Hemiptera

1. Fruit fly

Host range: Guava, Mango, Cucurbitaceous crops and other commercial fruits.

Damage symptoms: Damage is caused by maggots only and they feed on soft pulp, making the fruit unfit for human consumption. The infested fruits show small cavities with dark greenish punctures and when inter in pulp, the wriggling maggots are seen inside. The legless maggots, when full-grown, measure 8-9mm long and 1.5 mm across the posterior end and are yellow and opaque. The severe causes of fruits are rotting and dropping. Maximum activities of fruit flies are observed during august to the December reaching its peak during September. The flies are most active in gardens when the temperature ranges from 25-30°C and they become inactive below 20°C.

Biology: A female lays, an average 50 eggs under favourable conditions, inside the soft skin of fruits, 150-200 eggs are laid in one month. Egg hatch during 2-3 days in March-April and 1-2 days in the summer and 10 days during winter. Maggot pale cream, cylindrical, 5-8 mm in length, larval period 4-5 days. Maggot pupates in soil, pupal period 7-13 days. Adults are brown with greenish black thorax having yellow marking on abdomen. The life-cycle is completed in 2-13 weeks and many generations are completed in a year.

Biointensive Integrated Pest Management

1. Summer plough round the trees to expose of larvae and pupae stage of insects, pathogen and nematode to kill them by temperature and enemies.

2. Select deep, well levelled and well drained soil for this crop.

3. Collect and destroy of fallen fruits by burning them in the ground.

4. Use resistant rootstock and select insect and diseases free nursery plant.

5. Avoid flood and channel irrigation.

6. Harvest the fruits when slightly hard and green.

7. Monitor the fruit flies population in orchard by using methyl eugenol fruit fly traps 4-5/acre.

8. Make should be clean cultivation of orchard to avoid further development of fruit flies.

9. Adopt proper spacing, irrigation and nutrient management.

10. Avoid application of high nitrogenous fertilizers.

11. Ploughing the soil and application of *Beaveria bassiana* @ 5 kg/ha and *Metarrhizium anisopliae* @ 5 kg/ha to the soil underneath the tree canopy.

2. Tea mosquito bug

Host range: Guava, cashew, tea, neem and others

Damage symptoms: Both adults and nymphs feed on petioles tender shoots and leaf veins causing necrotic lesions, the develop blisters, scales rusty corky growth and scab formation on fruits widespread drying of shoots, inflorescence and flowers and shedding of fruits is witnessed. A single bug is capable of destroying two shoots per day. In case of severe attack causing defoliation, the shoots are killed and the plants appear like brooms. The bugs prefer a moist, warm atmosphere and that provide this microclimate are preferred by the pest.

Biology: A Female laid 30-35 eggs into epidermis of tender shoot, axis of inflorescence and tender fruits, eggs elongated and slightly curved with a pair of filaments. Eggs hatch during in 5-27 days, depending upon the prevailing temperature. A freshly emerged nymph is wingless and on account of its long appendages looks like a spider. It's completed during 15-16 days after passing through five moults. The adult is slender, 6-8 mm long agile and a good flier. It feeds at night on guava and several alternative host plants. During flight the adults may be carried long distances by the wind and get dispersed over wide area. Life cycle completed in 22-25 days.

Biointensive Integrated Pest Management

1. The pruning to regulate the shade to facilitate proper penetration of sunlight inside the canopy.

2. Collect nymphs and adults with hand net early in the morning or evening and destroy them.

3. Avoid application of high nitrogenous fertilizers.

4. Use resistant rootstock and select insect and diseases free nursery plant.

5. Avoid flood and channel irrigation.

6. Adopt proper spacing, irrigation and nutrient management.

7. Application neem seed kernel extract 5% are very affecting.

8. Application of bio-pesticide viz. *Verticillium leccani* culture @ 0.5% twice at fortnight interval during February-March and August-September.

3. Fruit borer

Damage symptoms: The caterpillars bore inside the developing fruits and feed on pulp and seeds just before the rind exhibiting round bore holes on fruit. Infested fruits are also attacked by bacteria and fungi, which ultimately fall off and that induce rot in the fruits producing a foul smell. There could be several caterpillars in a single fruit. The holes made in the fruits by larvae are usually visible. During severe infestation the loss caused by the pest is estimated to the tune of 90%.

Biology: The female lays shiny, white, oval shaped eggs laid singly on calyx of flowers and on small fruits. The eggs hatch in 7-10 days. The young larvae bore into the fruit and feed inside, larval full grown period 18-47 days, a single fruit may have about 6-7 caterpillar. They pupate period 7-34 days depending upon environmental conditions, inside fruit but occasionally outside even, attaching themselves to stalk of fruits. Pupa is dark brown in color. There are four generations completed in a year. The adult is about 2.5 cm across the wings and glossy violet in colour in case of male and violet brown in case of female.

Biointensive Integrated Pest Management

1. Collection and destruction of fallen infested fruits prevented build up of the pests.

2. Bagging of fruits before maturity will help in prevented damage.

3. The use of larval parasitoids of *Brachymeria euploeae*.

4. Destroy the infested fruits to minimize the damage.

5. Prune and destroy all the affected branches on timely.

4. Bark caterpillar

Damage symptoms: These are highly polyphagous species attacking a wide range of fruit, forest and avenue trees, especially the soft-wooded ones. The larvae bore into the trunk of branches make zig-zag shape galleries and contain wooden frass and excreta are the key symptom. It is very harmful to the tree as it feed on the bark and in this process it seriously injures the plant vessel through which nutrive plant–sap is transported within the plant's system, the result is that the tree growth and fruit bearing capacity are adversely affected. Sometimes the infested branches can dry up and in case of severe infestation, the whole tree may die. They remain hidden in the tunnel during day time, come out at night and feed on the bark. A severe infestation may result in the death of the attacked stem.

Biology: The female start eggs lying in may and June in cluster of 15-25 eggs each, in cracks and crevices under loose bark of trunk, stem or main branches, usually near the forks. A single female may eggs laid 2000 during summer. The eggs period 8-11 days and emerge larvae after 2-3 days bore inside. The larvae periods are 9-11 months to complete development. The larval period is quit long and the caterpillar continues its destructive activity from May-June to April the following year. The pupation starts from April onwards, inside the bark during 23-27 days. The pest only one generation is completed in a year. They become full grown by December where after they overwinter and pupate in April to emerge from the second week of May to first week of June. The adult is large sized moth with a wing-span of about 4 cm in the female and about 3 cm in the male. It is light grey to light brick-red in colour and with dark brown patches or dots.

Biointensive Integrated Pest Management

1. Cut and destroy the infested trunk and branches with larvae and pupae within.

2. The kill the caterpillars mechanically by inserting an iron spike into the tunnels. The mechanical destruction of the larvae, rather tedious on tall trees and in large area, is quite effective.

3. Injecting ethylene glycol and kerosene oil in the ratio of 1:3 into the tunnel by a syringe and then seal the opening of the tunnel with mud.

4. Dip a small piece of cotton in any of the fumigants, like chloroform or petrol or kerosene, introduce into the tunnel and seal the opening with clay or mud.

5. Keep the orchard clean and avoid overcrowding of trees.

5. Mealy bug

Damage symptoms: The bugs insect are found in large numbers sticking to leaves on ventral side, tender twigs and shoots. The bugs are both nymphs and adults suck cell sap from leaves, tender shoots and fruits them, that results in crinkling and yellowing of leaves. They also secret large quantity of honey dew which attract ants and favours development of sooty mould which interfere with photosynthetic activity. The severe infestation could kill the branches. They insect prefer dry weather.

Biology: A female lays 300-400 eggs of orange coloured in a loose cottony terminal egg-sac. Egg period 1-2 days and the young nymphs move away rapidly. The nymph period 35-40 days and the life cycle is completed in about 40 days. The adult bugs are shield shape, oval, yellowish green and measure 3mm in body length.

Biointensive Integrated Pest Management

1. Prune and destroy the affected parts at initial stage to prevent further infestation and build up.

2. Release lady bird beetle *Cryptoleamus montrouzieri* @ 10 beetles per plant.

3. Make orchard sanitation is very important for this pest.

4. Intermingling branch should be pruned and spacing trees at closed distance so that sunlight can reach through canopy from all the sides.

5. Ant colonies in the orchard should be destroyed as they are the carrier of mealy bugs to their feeding sites.

6. Release of the predator as *chrysoperla spp.* @ 2000/acre.

7. Ploughing the soil and application of *Beaveria bassiana* @ 5 kg/ha and *Metarrhizium anisopliae* @ 5kg/ha to the soil underneath the tree canopy

8. Application of bio-pesticide viz. *Verticillium leccani* culture @ 0.5% twice at fortnight interval during February-March and August-September.

6. Scale insect

Damage Symptoms: They suck sap from ventral side of leaves, tender shoots and occasionally from fruits. They cause leaf distortion and growth disturbance. The scale insect are found in large numbers sticking to leave on ventral side, tender twigs and shoots. The female feeds voraciously and also exude copious quantity of honeydew. The honeydew excreted by the scales encourages the development of sooty mould on foliage which interferes with photosynthesis activity of plant and spoils the market value of fruits.

Biology: The females are laid eggs beneath the body of mature female in a conspicuous egg-sac and later the female dies. The first instar nymph is the active dispersive phase responsible for starting new infestation. The adult scale are shield shaped, oval, yellowish green and measure 3 mm in body length.

Biointensive Integrated Pest Management

1. Prune and destroy the infested shoots at the initial stage of the infestation.

2. Use parasitoids and predators viz., *Coccophagus cowperi* and *C. bogoriensis* two species predators- *Scymnus coccivora* and *Crytolaemus montrouzieri* @ 20 per tree are most effective against this pest.

3. Intermingling branch should be pruned and spacing trees at closed distance so that sunlight can reach through canopy from all the sides.

4. Application of bio-pesticide viz. *Verticillium leccani* culture @ 0.5% twice at fortnight interval during February-March and August-September

5. Make orchard sanitation is very important for this pest.

6. Keep the orchard clean and avoid overcrowding of trees.

7. Adopt proper spacing, irrigation and nutrient management.

References

Atwal, A.S. and Daliwal, G.S. (2002). Agricultural Pest of South Asia and their Management, Kalyani Publication, pp. 301-303.

Azad Thakur N.S., Kalaishekhar, A., Ngachan, S.V., Saikia, K. Rahaman, Z. Sharma, S. (2009). Insect pest of crops in north east India. 360p. ICAR Research Complex for NEH Region, Umiam, Meghalaya.

Firake, D.M., Behere, G.T., Deshmukh, N.A., Firake, P.D. & Thakur azad, N.S. (2013). Recent Scenarioof Insect-pests of Guava in North East India and their Eco-friendly Management. *Indian Journal Hill Farming* 26(1): 55-57.

Kaul Virender, Shanker Uma and M.K. Khushu, (2009). Biointensive Integrated Pest Management in fruit crops Ecosystem, Integrated Pest Management: pp. 631-635.

Mathur, Y.K. and Upadhyay, K.D. (2005). A Text Book of Entomology, Aman Publishing House, pp. 301-305.

Srivastava, K.P. (2007). A Text book of Allied Entomology, Kalyani Publishers: 142-145.

Verma, L.R., Verma, A.K. and Gautam, D.C. (2004). Pest Management of Horticulture Crops: Principal and Practices, Asiatech Publisher Inc: 355-360.

9

Biointensive Integrated Pest Management in Mango

Gundappa Baradevanal and Kumarnag K.M.

Introduction

Agricultural pests cause considerable yield losses throughout the world. In ancient times, farmers had to manage pest problem to safeguard their basic needs, and as a response, the farmers use to practice and develop cultural and mechanical pest control strategies based on their field experience. Over a period of time, these practices have become a part of their production management system. Pest management practices began to change in later stages of the 18th century. Introduction of pesticides in agricultural ecosystems 1950's contributed substantially in raising agricultural productivity and they become an integral component of intensive agricultural ecosystems. At the same time, environmentalists raised their concerns on deleterious effects of pesticides on environment, human health and non-target organisms. Excessive use and abuse of pesticides leads to the problems like resistance, resurgence and residues (Beaument, 1993). One of the deleterious effects of over use of pesticides is the death of effective predator and parasitoids. This ultimately leads to the imbalance in the agroecosystem. Concerns about the negative effects of pesticides led to research and promotion of alternative pest control practices – Integrated Pest Control or simply IPC. This new concept called Integrated Pest Control (IPC) and later Integrated Pest Management (IPM) was stimulated by symposia organized by the Food and Agriculture Organization (FAO) of the United Nations in 1966 and the International Organization for Biological Control (IOBC) in 1967. IPM made a paradigm shift in the philosophy of pest control, from pest eradication to pest management. Instead of single tactic control, emphasis was placed on the use of a combination of available tactics in a compatible manner keep the pest below economic injury levels. Consequently, a more integrated approach of pest control was advocated considering the natural mortality factors. IPM is a complex system approach that comprises judicious use of cultural, physical, mechanical, biological, host plant resistance, regulatory and chemical methods.

Biointensive Integrated Pest Management (BIPM)

Biological control utilizing a population of natural enemies to seasonally or permanently suppress pests is not a new concept. The early success with the suppression of dreaded pests like red locust by Mynah in 1762, cottony cushion scale, *Quadraspidotus pernisicioci* with apple wooly aphid *Eriosoma lanigera* which nearly destroyed the citrus industry of California, was controlled by introduced predatory insects from Australia in the 1880s. Similarly, certain introductions of exotic natural enemies in India made around early 20th century gave success in many areas such as the suppression of wolly aphid of apple, *Eriosoma lanigerum* (Hausm.) by the North American hymenopterous parasites, *Aphelinus mali* (Hald.) in 1937 and of the polyphagous fluted scale *Icerya purchasi* Mask by Australian lady bird beetle predator *Rodolia cardinalis* (Muls) obtained from the USA in 1929 and from Egypt in 1930.The philosophy of Biointensive Integrated Pest Management Biointensive Integrated Pest Management is rely on conservation and enhancement of natural enemies and utilization of all compatible methods for achieving effective, economical and safe pest control/suppression. These methods are the most appropriate in perennial cropping systems and in protected cultivation.

This method of pest management reduces the pesticide pressure and enhances the farmers' income by fetching a higher remuneration for their produce and also ensures ecological sustainability. BIPM also opens new horizons to opt for a better choice, as biological control and use of bio-rational products, which are less toxic and only affect the target pest. It also includes biopesticides derived from microbials, parasitoids, predators, botanicals and all conventional non-chemical methods of pest management. In BIPM approach small area is intensively cultivated, using natural ingredients to rebuild and revitalize the soil health. Initially, it is more labor intensive than conventional approaches, therefore, it is better suited to the small farm areas. It lays emphasis on the use of predominantly indigenous or region specific cultivars, crop diversity, good soil health and water conditions which results in little or no pest problem. Consumers Union of the United States Department of Agriculture definese BIPM as, "A systems approach to pest management based on an understanding of pest ecology. It begins with steps to accurately diagnose the nature and source of pest problems, and then relies on a range of preventive tactics and biological controls to keep pest populations within acceptable limits. Reduced-risk pesticides are used if other tactics have not been adequately effective, as a last resort, and with care to minimize risks.

Mango (*Mangifera indica* L.) is an economically important fruit crop, popularly known as 'king of fruits'. This crop is ravaged by more than 492 species of insects and mites. About 45 per cent of these have been reported from India.

There are four to five key pests causing severe losses in mango orchards which includes fruit flies, stone weevils, mango hoppers, mealy bugs, scale insects and tree shoot borers. Several secondary pests may become major pests as result o certain aberration in cultural practices or indiscriminate and excessive use of insecticides against a key pest. In this chapter important pests of mango and their Biointensive management strategies were outlined.

Mango hoppers

Mango Hopper, *Amritodus atkinsoni, ldioscopus clypealis* and *I. nitidulus*

Its distribution can be seen throughout India and cause considerable yield loss. Nymphs and adults of hoppers puncture and suck the sap of tender parts, thereby reducing the vigour of plants and particularly destroying the inflorescence and causing fruit drop. Heavy puncturing and continuous draining of the sap causes curling and drying of the infested tissue. They also damage the crop by excreting a sweet sticky substance which facilitates the development of the sooty mould fungi which adversely affects the photosynthesis. The peak incidence of this pest was observed during March-April and it was lowest in December-January. Srivastava and Butani (1972) reported that hoppers have 2 or more broods in a year with two peak periods i.e. spring generation in February to April and summer generation from June to August. Patel *et al.* (1990) reported that population of *A. atkinsoni* starts increasing with the beginning of the flowering season in the month of January to June and adult population build up was seen from March onwards and a gradual fall from July onwards was observed.

Biointensive management strategies for mango hoppers

Studies on the comparative efficacy of essential oils of *Mentha spicata, Cymbopogan winterianus, Cymbopogon fiexuosus, Mentha citrata, Mentha arvensis and* palmorosa oil under field condition against mango hoppers during 1994 and 1995 found that 0.125% of *M. spicata*, lemongrass oil, citronella oil, *M. arvensis* and *M. citrata* at 0.25% were found efficacious (Srivastava, 1995).

An oil-based and a kernel based neem concentrates proved effective against *A. atkinsoni* and *I. nitidulus*. The oil based concentrate (1%) and the kernel based concentrate (0.2%) caused 100% and 66.6% mortality, respectively, of nymphs of *I. nitidulus* in laboratory. The efficacy of the formulations was also determined in the field against *I. nitidulus*, while oil based concentrate, killed the percentage of nymphs as in the laboratory (Srivastava *et al.*, 1993).

Srivastav *et al.* (1980) recorded 12 species of spiders belonging to eight families namely *Phiddiphus* sp., *Rhene indicus, Marpisa* sp., *oxyopes shweta, Cyrtophora* sp., *C. cicatrasa, Araneus sinhagadensis, Chieracantium donicli, Linylia* sp. *Stegodyphes sarsihorum, Uloborus* sp., *Hersilia sarigryi* and *Theridion indica.* They also observed *Coccinella septumpunctata, C. transversalis* and *Menochilus sexmaculatus* preying upon *I. clypealis.*

Two species of mantids, and two species of neuroptera i.e *Mallada boninensis* and *Chrysopa lacciperda* were found preying upon nymphs adults of *I. nitidulus* besides *Bochartia* sp., a mite preadating the nymphs in Lucknow. Larvae of these neuropterans were found predating on nymphs of hoppers in flowering season from February to May as well as at the time of emergence of new flush in July to September. Eggs and larval stages of these predators were found mostly on panicles and shoots and puparium on the bark.

Isyndus heros Fabracius, *Epipyrops fulginosa* Tames, *Pipunculus annulifemer* Brundli and *Halictophagus indicus* Bohart were recorded on mango hoppers (Ramachandra Roa 1930; Wagle, 1934; Rehman, 1939). The parasitized hoppers are very sluggish and with bulging abdomen. The seek dark and shady places and are found adhering to the under surface of the leaf. The full grown maggots come out of the dorsal surface of the abdomen and drop down pupate in the soil. The hoppers become sluggish. Another species parasitizing the adult hopper is *Pyrilloxenos paracompactus* Pierce. The female of this species is wingless and buried in the abdomen of the host. The larvae, after hatching come out of the anterior portion of the female which is projected outside the body of the host. These larvae are small active, fish like creatures of plae yellow colour. The male pupae are found buried on the pleura-dorsal side of the body of the host. They appear as small, globular bodies. The death of the hopper is sure in case of parasitization. It has been observed that 30 per cent of the parasitized hoppers are *A.atkinsoni* and 15 per cent of *I.clypealis.* Generally one and occasionally 2 or 3 parasites, may be found in a hopper. Larvae of *Epipyrops fulginosa* Toms (Epipyropidae:Lepidoptera) also found parasitizing the hoppers. They were found attached to the side of thorax. The larvae are of cream colour. Pupation takes place in white silken cocoons on the ventral surface of the leaf. Egg parasites *Aprostocetus* sp., *Gonatocerus* sp., *Polynema* sp. and *Tetrastichus* speices were recorded on *I. nitidulus* and *A. atkinsoni.* The extent of parasitization was 5-10 per cent as observed in the eggs of *I. nitidulus* at the time of epidemic outbreak of the this species in 1987 in Lucknow, India (Srivastav *et al.,* 1987).

Entomogenous fungus, *Verticillium lecanii* and *Beauveria bassiana* were found pathogenic to *I. clypealis* (Srivastava and Tondon, 1986). Pathogenicity tests, conducted in the laboratory, gave cent per cent mortality within 5 days when the fungal spores were sprayed and within 3 days when the hoppers were allowed to crawl over the dense culture of the fungus, *V. lacanii.* Observations recorded in different orchards revealed that number of number of parasitized hoppers varied from 1-29 per leaf with an average of 9.33 hopper per leaf. Natural occurrence of fungus, *Isaria tax* on *I. clypealis* and *A. atkinsoni* was also recorded from orchards in Tripura state in India and its pathogenicity was confirmed in the laboratory tests (Kumar *et al.*, 1985). Besides, *Hirsutella* sp. was also recorded on *I. clypealis* (Choudhary *et al.*, 2012).

Mango mealybug, *Drosicha mangiferae* Green

Mango mealybug, *Drosicha mangiferae* Green is a polyphagous insect pest, feeds on 71 plant species (Tandon *et al.*, 1978; Srivastava, 1997). Infestation due to this pest leads to significant loss in size and weight of fresh mango fruits and causes yield loss up to 80 per cent (Karar, 2010; Karar *et al.*, 2012). Mango mealybug is considered as serious pest in India, Bangladesh, Pakistan and China. In India, infestation of this pest is quite serious in Punjab, Uttar Pradesh, Bihar and Delhi (Srivastava, 1997). The pest being in polyphagous in nature has a very wide host spectrum which includes fruit trees, forest trees, ornamental plants, vegetable plants, weeds and grasses etc. So far it has been recorded on 71 plant species (Srivastava, 1997).

The nymphs and adults are damaging stages, they suck sap from inflorescence, tender leaves, shoots and fruit peduncles. As a result, the affected inflorescences are shriveled and get dried. Severe infestation affects the fruit set and causes fruit drop. They secrete honey dew over infested part, on which sooty mould develops. Due to the growth of sooty mould on the leaves, photosynthetic activity is affected (Karar, 2010a).

The mango mealybug (*D. mangiferae*) nymphs started to hatch out at end of December or beginning of January. A single female lays up to 400-500 eggs. The duration of 1st instar vary from 45-71 days; second 18-38 days; third instar for female 15-26 days. On mango the total duration is 77-135 days for female and 67-119 days for male. The copulation time of male with female was 4-10 minutes and the ratio of males to females was 1:19 (Ishaq *et al.*, 2004; Karar, 2010b; Haseeb *et al.*, 2003).

Population dynamics (Yadav *et al.*, 2004; Kumar, 2009) of mango mealybug and impact of weather variations on the incidence of mango mealybug have been studied. Population of mealy bugs starts appearing during last week of December. The population of mealy bug then increase at weekly interval till the middle of March to first week of April. First instar mealy bug can be seen on panicles in the beginning *i.e* middle of February, there after second instar nymphs start moulting and crawling on tree during next fortnight. The second instar nymphs also start moulting from mid March, which again moulted to fourth instar nymphs in mid April and further they are being called as full grown up mealy bug with developed ovisac in last week of April and onwards. The population of mealy bug seen coming down the tree to the ground with its ovisac in last week of April which continued later till the date and pupate inside the soil (Srivastava, 1997).

Biointensive management for mango mealybug

Important Biointensive management options for mango mealybug includes ploughing of orchard, banding of tree trunk with alkathene above ground level (Bindra *et al.*, 1970; Srivastava, 1980), application of entomopathogenic fungi, *Beauveria bassiana* (Haseeb and Srivastava, 2003; Haseeb *et al.*, 2003) and spraying of botanicals (Tandon and Lal, 1980a). The efficiency of cultural methods can be augmented further by the application of physical barriers like sticky bands. One sticky band per tree, applied during the month of December was significantly better in restricting the mealy bug movement upwards (Sandhu *et al.*, 1980). The grease band is effective for checking the migration of mango mealybug. Similarly, polyethylene sheeting is an effective barrier to prevent the upward moving nymphs of mango mealybug and was much cheaper, easily accessible and practical. Whereas, alkathene sheeting was more effective than polyethylene against upward crawling nymphs. Double girdle band of alkathane sheeting was the more effective than single girdle alkathene bands. Funnel Type Trap was also found effective barrier for mango mealybug nymphs and also worked for collecting the egg carrying female.

Use of predator named *Sumnius renardi*, which has the ability to congregate inside the tree bark and other physical cavities. Burlap wraps were used to provide protection to congregating predator, which resulted in the survival of a large population of this predator. The predator feeds on climbing nymphs of the mealy bug and effectively reduces the population of climbing mealy bugs by 84.09%. Integrated biointensive strategies for effective management of mango mealy bug inclues ploughing of orchard in November-December, banding of tree trunk with alkathene (400g), 25 cms wide 30cms above ground level and application of *Beauveria bassiana* (2 g/l) or five per cent NSKE in second week of December around tree trunk and conservation of biocontrol agents *viz.* fungus, *Beauveria bassiana,* predators. *Menochilus sexmaculatus, Rodolia fumida and Sumnius renardi.* Bokonon-ganta and Neuenschwander (2010) reported that mango mealy bug *Rastrococcus invadens* Williams (Homoptera: Pseudococcidae) incidence was controlled effectively from 31% to 17.5% by introducing exotic parasitoid *Gyranusoidea tebygi* in Benin.

Mango Midges

Midges (Diptera: Cecidomyiidae) are considered as important pests of mango. About 16 species of gall midges attack mango in Asia (Harris and schreiner, 1992; Pena, 2002). Mango midge (*Erosomya indica*) has gained much attention in recent past as it has become major pest in all mango growing areas of the world (Ahmad *et al.,* 2005). The mango gall midge or mango blister midge (*Erosomya mangiferae* Felt.) is a major pest destroying flowers and upto 70 per cent of fruit set. Similarly the leaf gall midge (*Procantarinia mattieiana*) is a serious pest of mango in Oman. It is feared that heavily infested mango trees may produce few inflorescence, resulting in reduced yields of mango fruits. Galled leaves remaining on trees are known to provide reservoirs of anthracnose inoculums. This pest is distributed in India, Indonesia, Kenya, Mauritius, Oman, Reunion, South Africa and United Arab Emirates. In India 12 species midges representing three genera were known to produce different types galls on mango leaves. Recently, this pest has become very serious in certain pockets of Uttar Pradesh, causing serious damage to mango crop by attacking both the inflorescence and small fruits (Srivastava, 1997; Rajkumar *et al.,* 2013).

The midge infests newly emerged panicle by ovipositing at bud burst stage, and the first instar maggots bore into the growing panicle. Infested panicles have characteristic right angled bend, with an exit hole, from which last instar maggots emerge to pupate in the soil. The second generation then infests on very young fruits, which eventually drop before the marble stage (Pena *et al.,* 1998). The midge has four larval instars, and field cage traps showed emergence of adults to be in the afternoon. Infestation was noticed at bud burst stages, at fruit set and on tender leaves of new flushes. The population of the insect is less in the month of January, whereas the infestation increases during the month of Febraury and March; then in April population decreases (Rajkumar *et al.*, 2013).

Biointensive management for mango midges

As the larvae pupate in the soil, ploughing of the orchards expose pupating as well as diapausing larvae to sun heat which kills them. In a survey of parasitoids of cecidomyiid pests of mango in India, Grover (1986) found that *Platygaster* sp., *Systasis* sp. and *Euplemus* sp. were associated with *Dasineura* sp. and *Tetrastychus* sp. were associated with *Eryosomya indica*. An external parasitoid, the pteromalid *Pirens* sp., was found attacking *Procystiphora mangiferae* (Felt.). Predators of the cecidomyiid included *Formica* sp.; *Oecophila* spp. and *Camponatus* spp. Integrated Biointensive strategies for mango midges includes deep ploughing of orchard in November to expose pupae conserving natural enemies.

Shoot gall psylla, *Apsylla cistellata* Buckton

Shoot gall psylla, *Apsylla cistellata* is a monophagous pest of mango in northern India. Mango shoot gall psylla, *Apsylla cistellata* Buckton is one of the most noxious pests. Many biological investigations on *A. cistellata* have been reported mostly from the northern plains of India viz., Uttar Pradesh, Bihar and Tarai regions of northern India. (Monobrullah *et al.* 1998; Singh, 2003). In recent years this pest incidence is increasing towards Lucknow (Uttar Pradesh) which is major mango growing area (Gundappa *et al.*, 2014). According to Singh & Misra (1978) the female insect inserts the eggs in the midrib of the leaf at emergence of new flush in March-April in two parallel rows. The freshly laid eggs are whitish translucent oval in shape with its tip partly exposed, which is characteristic. Eggs hatch either in mid-September or early October, approximately 200 days after oviposition. Nymphal period includes five instars and the development into adults takes about 140 days. Gravid females never oviposit on the leaves of seedlings; they prefer tender leaves of mature plants ready to bear flowers and fruits (Singh, 2003). Shoot gall psylla is having most peculiar feature of life cycle is the feeding of nymphs inside the leaf midrib without proper egg-hatch. Feeding of nymphs and subsequently secretion of certain chemicals through the 'triggers' gall development, although the first-instar nymph, without proper egg hatch, remains partly within the egg shell and feeds on the same leaf where the adult female oviposited (Singh *et al.* 1975). Almost everyone who studied these galls and the psylloid indicates that galls develop through the modification of vegetative axillary buds as the first-instar nymphs feed on the leaves and only the second-instar nymphs migrate to the already organized galls. The feeding effect of multiple neonate nymphs (from an egg cluster of 75-150/leaf) possibly induces the modification of adjacent vegetative buds into galls, in an approximate period of 30 days. An increase in endogenous auxin levels and a decrease in total phenols and contents of auxin precursors (e.g., tyrosine and tryptophan) in the buds of *M. indica* due to the pest has been demonstrated (Singh, 2000). The gall formation caused by this pest is only after the tree starts flowering and fruiting, which directly interfere with the formation of inflorescence and thus adversely affect the yield of mango crop. In due course of time, infested twigs dry showing die-back symptoms.

Biointensive management for mango shoot gall paylla

Practice of removal of eggs bearing leaves from a shoot leads to 95.57 per cent decrease in number of shoot galls. This practice may follow to reduce pest incidence (Kumar *et al.*, 2007a). Pruning of affected shoots helps in effective management of shoot gall psylla. One parasite *Inostemma apsyllae* (Platygastiridae: *Hymenoptera*) and two predators *Buccha pulchirifrons*

(Syrphidae: *Diptera*) and *Micronum timidus* (Hemorobidae: *Nueroptera*) were recorded on this pest.

Mango leaf webber, *Orthaga euadrusalis* Walker

The mango leaf webber has become a major limiting factor in mango production in Uttar Pradesh, Bihar and other parts of north India like Malda district of West Bengal (Srivastava, 1997; Rajkumar *et al.*, 2013). This pest incidence was also observed in south India in severe form (Kavitha *et al.*, 2005). The insect recorded on mango trees in India are *Orthaga euadrusalis* Walker, *Orthaga exvinacea* Hampson and *O. mangiferae* Mishra. Verghese (1998) reported that severe infestation of mango leaf webber results in complete failure of flowering.

The caterpillars on hatching reached to tender leaves nearby and fed gregariously on leaf chlorophyll by scraping the leaf surface. In young stages, the caterpillars webbed two to three leaves together and fed on them by cutting the leaves from edges towards the midrib leaving behind the network of veins. In grown up stages, the larvae were found feeding voraciously and webbing the shoots and leaves together. The leaves loosened from their stalks, often detached but remained entangled in webs on the tree. Numerous dried bunches of shoots and leaves are clearly visible from a distance on severely attacked mango tree (Rafeeq and Ranjini, 2011).

Adult moths are medium sized and sombre coloured, lay their eggs on leaves. The eggs are greenish dull in colour and hatching takes place in 4-7 days. First instar larvae feed on leaf chlorophyll and, from second instar onwards they start webbing the leaves and feed on entire leaf, leaving behind the midrib and veins. The webbed leaves give a small tent appearance, so it is also popularly called as 'tent caterpillar'. Full grown caterpillar measures between 2.5 and 3 cm, and its colour is of a brownish blue with whitish striations dorsally. Its head is symmetrically variegated with brown and white spots and markings, there are four whitish longitudinal striations dorsally alternated in between by three pale olive green bands. Laterally there are two brown to black striations and dorsally, two rows of symmetrical dots run longitudinally along the body. The caterpillar is smooth having spars hair-like white setae. A larva undergoes 5 instars, each instar lasts in 2-3 days duration in July. Larval period varies between 15-33 days. The caterpillars when disturb fall with a sudden jerk. They pupate on the leaves usually within the web, but the last generation caterpillar in December-January secrete thread by which they hang and descend on to the ground for pupating in the soil. They spin a cocoon, on which soil gets adhered. Pupae are dark blackish on maturity measuring 15 mm in length and 4 mm in width. The suitable temperature for pupation was 35°C and pupal period lasts from 4-5 days. At 25°C the pupal period lasts from 16-18 days. The hibernating larvae pupate in March and adult emergence in April end (Haseeb *et al.*, 2000; Singh, 2002; Beria *et al.*, 2008). The webber infestation begins from June and continues upto December. There are five generations in year (Bhatia and Gupta, 2002). Lakshmi *et al.* (2011) reported that the peak incidence of leaf webber was observed during 44th SMW at Hyderabad, Andhra Pradesh.

Biointensive management for mango leaf webber

Brachymeria lasus (Walker), *Tetrastichus* sp., *Pediobius bruchiada* (Rondanii), *Hormius* sp., *Cathartoides* sp. were recorded in Uttar Pradesh (Tandon and Srivastava, 1980). *Asperigillus flavus*, *Beauverria bassiana* and *Serretia marcescense* were recorded from Malihabad (Singh, 1993a). *Gonoizus* sp. was recorded as an important larval parasitoid in Kerala (George and Abdurrehaman, 1985). Mechanical removal of leaf webs infested by leaf webber by leaf web removing device and bur them. Ploughing of orchard done earlier for mealy bug control will check its population.

Mango Nut (Seed) Weevil, *Sternochetus mangiferae*

The mango nut (seed) weevil (*Sternochetus mangiferae*) belongs to the Family Curculinoidae of Order Coleoptera. It is restricted to mango (*Mangifera indica)* and affects the plant at fruiting stage. It infests fruits and seeds (nuts). It is reported in India from union territory of Andaman & Nicobar Islands and states of Andhra Pradesh, Assam, Karnataka, Kerala, Maharashtra, Manipur, Orissa, Tamil Nadu, Tripura and West Bengal. The pest is reported to occur in Australia (Australian Northern Territory, New South Wales and Queensland).

Eggs are laid on the epicarp of partially developed fruits or under the rind of ripening fruits. Newly emerged grubs bore through the pulp, feed on seed coat and later cause damage to cotyledons. Pupation takes place inside the seed. The pulp adjacent to the affected stone is seen discoloured when the fruit is cut open.

Adults of *S. mangiferae* feed on the leaves and tender shoots of mangoes during March and April. They are nocturnal, fly readily and usually feed, mate and oviposit at dusk. After emergence, adults enter a diapause, which varies in duration with the geographic range. For example, in southern India, all adults emerging during June enter a diapause from July until late February of the following year. Adults are capable of surviving long, unfavourable periods. During non-fruiting periods, weevils diapause under loose bark on mango tree trunks and in branch terminals, or in crevices near mango trees. A few adults live through two seasons with a diapause period in between.

Biointensive management for mango stone and pulp weevils

A very few natural enemies of mango pulp and stone weevils had so far been reported. The mango stone weevil, *S. mangiferae* has few natural enemies (Hansen, 1993). According to him, the adults may be susceptible to predation by ants, rodents, lizards and birds. Parasitoids are unknown probably because of the secretive behavior of the most life stages (Hansen, 1993; Schoeman, 1987). A baculovirus infecting mango nut weevil in India was identified (Shukla *et al.*, 1984). The virus caused loss of appetite, sluggishness, browning of the integument and milkiness of the haemolymph. In India, apathogen, Beauveria bassiana (Balsamo) Vuillemin was found. The natural occurrence of *B. bassiana* on mango nut weevil was less than 1 per cent (Verghese *et al.*, 2002). The weaver ant, *Oecophylla smaragdina* (Fabricius) is an effective biocontrol agent of *S. mangiferae* in the northern territory of Australia (Peng and Christian, 2007). Voute (1935) reported that in China and Indonesia, the red ant, *O.smaragdina* preys upon the adults of *S. gravis*. Dey and Pandey (1988) observed that two species of ants, viz., *Oecophylla smaragdina* and *Camponotus* sp. abundant on mango trees were found to be disturbing mattingand egg laying of *S. gravis*. Another species of ant. *Monomorium* sp. was found to devour adults of *S. gravis* by cutting them into pieces. They also killed exposed grubs. Hibernating adults, both in field and laboratory were frequently infected by *Aspergillus* sp. The larvae, pupae and hibernating adults were also frequently parasitized by ecotoparasitic mite *Rhizoglyphus* sp.

Mango Shoot borer, *Chlumetia transversa*

Mango shoot borers have been reported to cause serious damage to mango trees. The caterpillars are reported to bore young mango shoots in mango growing countries of the world. The species reported from India are *Gateseclarkeana (Argyroploce) erotias* Meyrick. *Anarsia melanoplecta* Meyrick, *Anarsia lineatella* Zeller, *Chelaria spathota* Meyrick (Fletcher, 1916) and *Chlumetia transeversa* walker, *Chlumetia alternans* Moore and *Dudua (Platypelpa) aprobola* (Meyrick). Besides these, *Chlumetia brevisigna* Holloway has been reported on mango from Japan. *Chlumetia transeversa* has been also reported from Bangladesh, China, Java and Philippines.In India this pest has been reported to cause serious damage to mango shoots in Rajasthan and Uttar Pradesh (Srivastava, 1997).The attack is noticed during the period when there is a new flesh on the trees and saplings. The young saplings are attacked during the earlier part of April with the commencement of hot winds, the caterpillars pupate and activate till the onset of monsoon when they were again found active to attack the new flush. The larvae, of this bore in to the shoot which dies and does not flower subsequently. The larva eats the young leaves and inflorescence.

The trees which have been top worked, have the marked damage of this insect. This pest appears in June-July and its incidence will be high during July-October after which it declined (Bhole *et al.*, 1987). The adult female lays eggs singly on the tender leaves, caterpillar after hatching from the egg measures 3 mm in length and is pale white in colour with black head and prominent legs. The full grown caterpillar is pink with dirty white spots dorsally and ventrally and is pale white. The caterpillar takes 10-12 days to mature and then it leaves the tunnel and enters into the cracks and crevices of bark of the tree, dried and malformed panicles and also in the soil for pupation. A pupa undergoes diapause till the next monsoon. After onset of mansoon pupae will take 15 to 18 days to adult emergence. Moth leaves upto 15 days. This insect completes four generations in a year (Chahal and Singh, 1977).

The adult female lays eggs singly on the tender leaves and hatch within 2-3 days. The newly hatched larva bores into midribs of these leaves and feeds therein for 2-3 days and, thereafter it comes out and bores in to the tender shoots. They make tunnels downwards upto 100 mm to 150 mm in length and expelled out excreta through the entrance hole and the shoot become hallow. The affected shoots show dropping of leaves and give a wilting look.

Biointensive Management for mango shoot borer

Attacked shoots should be clipped off and destroyed.

Fruit flies, *Bactrocera dorsalis* Hendel and *B. zonata*

Fruit flies are of major economic importance because many representatives of this family attack and severely damage important fruit crops, especially mangos, in tropical regions. Fruit flies are the serious pest of mango and cause economic losses. The importance of fruit flies has gone very high due to the major constraint in the export of fresh mango fruits to foreign counties. Kapoor (1970) listed 128 species of fruit flies and out of these eight are found infesting mango fruits in India. They are *Bactrocera dorsalis, B.zonata, B. correcta, B. zonata, B. diversa, B. cucurbitae, B. hageni,* and *B. tau.* The insect is distributed throughout India; in the North it overwinters in pupal dormancy but in the South it is active throughout the year.

A female fly punctures the epicarp of mature fruits with ovipositor, insert eggs in clusters into mesocarp. This leads to sting marks and bruising to the fruit skin constitute the external damage that later turn to brownish rotten patches. Following hatching, larvae feed on the pulp of fruits; infested fruits though appear normal from outside, eventually drop. The females flies lay clusters of 6-10 eggs just under the skin of the fruit. After 1-2 days larvae hatch from the eggs and take 6-8 days to mature. Larvae feed upon the pulp of fruit. The larvae pupate in soil (5-10 cm) and flies start emerging from April onwards with maximum population during May to July which coincides with fruit maturity. The adults emerge after 10-12 days and may live for a few months.

Apart from mango this pest also feeds on guava, peach apricot cherry, pear, sapota ber, citrus and other plants, totaling more than 250 hosts. The emergence of fruit fly starts from April onwards and the maximum population is recorded during May-July which coincides with fruit maturity. The population declines slowly from September-December.

Biointensive pest management for fruit flies

Avoid infestation of fruit flies by early harvesting of mature fruits. To prevent the carry over of the pest, collect and destroy all fallen infested fruits twice in a week. Plough round the trees during winter to expose and kill the pupae. Gamma irradiation of fruits with dose of 600 GY was found effective in killing the immature stages of fruit flies. Vapour heat treatment of fruits at 48°C for 1 hour also found effective.

The parasitoids associated with this pest are *Opius compensates* Silvestri, *O. persulcatus* Silvestri, *Biosleres arisanus* (Sonam), *O. incises* Silvestri and *O. manii* (Braconidae); *Spalangia philiphinensis* Mill., *S.afra*, *S. stomoxysine* Gir. and *S. grotiuse* Gir. (Pteromalidae); *Dirhinus giffardi* Silvestri (Chalcidiae); *Pachycrepoideus dubiers* Ashmead and *Trybliographa daci* Weld (Eucoilidae).

Scale insects (Coccids)

Over 70 scale insects belonging to coccid family Diaspididae, Flatidae and Lecciferidae are reported attacking mango. Among all *Pulvinaria* spp., *Ceroplastes* spp. *Aspidiotus* spp. And *Rastrococcus* spp. occasionally causes severe damage to mango tree. Among these *Chloropulvinaria polygonata* becomes a serious pest in western Uttar Pradesh. *Rastrococcus iceryoides* and *Aspidiotus destructor* also some time cause severe infestation (Srivastava, 1997). The insects of coccids are very minute in size and the biggest size of this group is below 2.5 cm. all the adults have a covering of waxy, horny, glassy or resinous secretion or some sort of powdery white materials. The adult females are wingless, while the males are winged. The female leads a stationary life on the plant parts except the mealy bugs which are able to move about slowly all their life. They feed the sap of the plants by well developed thread like sucking tube. The adult female produces numerous minute eggs which are found in a pouch. The time when the eggs inside the mother become mature and nearing hatching time, the under surface soft portion gradually shrinks and female dies away thereby giving opportunity to the eggs to hatch. From the eggs, tiny larvae hatch out which possess 6 legs, a pair of feelers, a well developed feeding tube and minute eyes. These larvae crawl to the tender parts of the plants and shortly attach themselves at a spot. The periodical moulting takes place and the larva loses its original form and becomes a small footless mass covered over the scale. It is difficult to differentiate male and female larva at this stage but after 1 to 2 moths the male insects gradually emerge as winged forms. These males mate with female and die soon. The female fix themselves with the place for entire life span. They loose their feet, feelers etc. and grow by feeding the sap and secrete waxy covering over the body called puparium. Like other hemipterous insects these insects also reproduce parthinogenitically. The main period of activity of this insect is summer when they breed rapidly, although in cold and rainy season breeding is at reduced rate.

Biointensive management for mango scale insects

Pruning of the affected plant parts such as leaves, twigs, and branches helps in reducing the infestation clean planting material would also minimize the scale population. Due the presence of the thick waxy layer on the body of the scale, it becomes very difficult to control the insects by chemical insecticides but at the same time they are found heavily parasitized by the parasites and consumed by the predators. It is now essential to supplement the introduction of bioagents by filing a gap in natural enemy spectrum with exotic species of bioagent. To permit the parasite population to build up, it is necessary to keep unsprayed alternate plants or a row in the orchard.

A large number of scale insects have been recorded on mango. Among these, *Pulvinaria* spp., *Ceroplastes* spp., *Aspidiotus* spp. and *Rasrococcus* spp. causes sever damage to mango tree. In case of *R. iceryoides*, as high as 42% parasitization was noticed and the major parasites are *Anagyrus pseudococci*, and *Promascidia unfascitiventris*. Among predators *Cryptolaemus montrouzieri* is very efficient predator besides other coccinellids. Larvae of midge predator *Coccodiplosis* sp. were noticed feeding veraciously on this coccid. All the stages viz. egg, larva, and pupa are found among the colony. This species is fast breeder, has short life cycle and is an active predator. In case of *A. destructer*, the parasite *Compriella aspidiotiphaga*, *C. bifasciata* and *Chartocerus* spp. Are very important ones. The important predators are *Chilochorus nigritus* and *Scymnus* sp.In case of *Chloropulvinaria polyginata*, the parasites *Anagyrus* sp. and *Anicetus annulatus* are important (Tandon and Lal, 1978; Mani *et al.*, 1995).

Daneel and Dreyer (1998) reported that two biological control agents, an *Aphytis* sp., a parasitoid and *Cybocephalus binotatus*, a nitidulid beetle, were found very effective in reducing the mango scale, *Aulacaspis tubercularis* population under field condition in Africa.

Mango thrips

There are twenty species of thrips reported attacking mango viz., *Scelenothrips rubrocinctus* (Giard), *Thilibothrips inquilims* sp. n, *Scirtothrips dorsalis* Hood *Taeniothrips varicornis*, *Rhiphiphorothrips cruentatus* Hood, *Thrips hawaiiensis* (Morg), *Scirtothrips mangiferae* Priesner, *Rhamphothrips pandens* sp.n, *Thrips palmi* Karney *Frankliniella occidentalis* Pergande, *Thrips tenellus*, *scirtothrips aurantii*, *megalothrips destalis*, *Haplothrips ceylanicus* schmutz, *H. ganglbaueri* schmutz, *H. tenuipennis Aeolothrips collaris* Priesner, *Anophothrips sudenesis* Trybom, *Ramaswamiahiella subnudulu* karny, *Neoheegeria mangiferae* (Priesner) and *Thilibothrips inquilinus* sp.n. It is an emerging pest attacks mango during summer. In case of severe infestation it found to attack immature fruits of mango by lacerate and sucking the juice in addition to leaves, buds, and flowers. The affected fruits show rusty appearance on their skin. The curling up of leaves and wilting of inflorescence are common symptoms. This pest targets new leaves, especially in rejuvenated trees it is observed as severe and the entire new growth will be affected if not treated properly.

Selenothrips rubrocinctus

This is called as red banded thrips, also called as cocoa thrips but its main host is mango. This belongs to the family Thripidae. It also infests avocado, pear, cashew guava and cocoa. This is a very serious pest in mango nurseries and is rarely found damaging the big trees. They first pierce the epidermis and then scrape out the leaf tissue below, leaving a minute spot which becomes brown. Due to its damage leaves of the growing tips blacken, curl and drop off. Sometimes defoliation may take place. This pest is distributed in parts of Asia, Africa, Australia, the pacific Islands, the America and the West Indies.

The eggs of this pest are kidney shaped about 0.25 mm long and are inserted into the leaf tissue by the female; incubation period lasts from 12-18 days. The nymphal stages are yellow with a bright red band round the base of the abdomen. The full grown nymphs are of 1 mm length. A nymph feeds on the underside of the leaves. The total nymphal period lasts for 6-18 days. The pupal stages are passed on a sheltered spot in the curls of the leaves. Both pre-pupal and pupal stages resemble nymphs but differ in well developed wing pads. Pupal period lasts for 3-6 days. Adult thrips are dark brown and just over 1 mm long.

Rhipiphorothrips cruentatus

This is commonly called as grapevine thrips. This is highly polyphagous and feds on custard apple, cashew nut, almond, guava, mango and pomegranate. The pest is active throught the year. This pest cause injury by puncturing and sucking sap from epidermis of leaves and affected area turned dark or developed scars. The leaves become blackened on their growing points, curled and finally fell down. In case of high infestations, complete defoliation occurs. The punctured places serve as source of entry of fungal attack.

The adult female lays in slits with inserting ovipositor in a leaf. Generally upto 50 eggs are laid by a single adult female. The incubation period lasts for about 4-6 days. There are 4 nymphal instars, third is called pre-pupa and fourth is known as pupa. These stages normally develop in 10-25 days except in winter when the pupal stage hibernates in soil. The total life cycle varies from 14-33 days. There are 5-8 generations in a year.

Scirtothrips dorsalis

This causes heavy losses to chillies crop. This species also infests castor, tea, cotton, tomato, sunflower, mango, Acacia arabica, *citrus* sp., *Albizia odorotasimmi, melia azadiracta, Mimosa pudica, Vigna mungo, punica granatum, tamarindus indica, vitis viniofera and Ziziphus mauritina*. They occur in such a large number they suck up the sap from tender regions and cause the leaves shrivel in severe cases there is a malformation of leaves, buds and fruits. This species reproduce sexually and parthenogenitically, ovipositing within the tissues of tender leaves. The adult female lays 48-50 eggs. The life cycle is completed in 15-20 days depending on weather. The female male ratio is 6:1. There are as many as 25 generations in a year.

Thrips hawaiiensis

These are important flower inhabiting thrips and cause severe damage at times. They cause brownish streaks in flowers and seed setting is affected. This thrip was recorded by Srivastav and Tandon (1982) for the first time damaging on the mango inflorescence from India. This pest was recorded on Chinese cabbage, turnip, rose, apple, bitter gourd, citrus and wheat. Peak incidence of this pest can be observed during March and April. White eggs are laid in the notches of epidermis of leaves. The incubation period lasts from 4-10 days nymph moults twice in about 5 days. Pupation occurs in soil and takes 4-7 days. The adult is small brown thrip with darker transverse bands across the thorax and abdomen. They are 1 mm in long.

Biointensive management for thrips

Monitor for thrips infestation by placing sticky traps at regular intervals. Neem based pesticides control young nymphs effectively, inhibit growth of older nymphs and reduce the egg-laying ability of adults. Promoting natural enemies that include predatory thrips, predatory mites (e.g. *Amblyseius* spp.) anthocorid bugs or minute pirate bugs (*Orius* spp.), ground beetles, lacewings, hoverflies, and spiders. Peng and Christian (2004) evaluated the efficacy of The weaver ant, *Oecophylla smaragdina* (Hymenoptera: Formicidae) against mango red-banded thrips, *Selenothrips rubrocinctus* (Thysanoptera: Thripidae) incidence in mango Australia and it was found very effective in reducing the thrips incidence.

References

Ahmad, W., Nawaz, M.A., Saleem, A. & Asim, M. (2005). Incidence of amngo midge and its control in mango growing countries of the world. In *II International conference on Mango and datepalm 231*, pp. 98-102.

Beaumont, P. (1993). *Pesticides, Policy and People*. Pesticide Trust, London.

Beria, N.N., Acharya, M.F. and Kapadia, M.N. (2008). Biology of Mango Leaf Webber, *Orthaga exvinacea. Annals of Plant Protection Sciences* 16: 218-219.

Bhatia, R. and Gupta, D. (2002). Incidence and the control of the mango leaf webber, *Orthaga euadrusalis* walker (Pyralidae: *Lepidoptera*) in Himachal Pradesh. *Agricultural Science Digest* 22: 111-113.

Bindra O.S., Varma G.C., Sandhu G.S. (1970). Studies on the relative efficacy of banding materials for the control of mango mealy bug, *Drosicha stebbingi* (Green). *J Res Punjab Agric Univ* 7(4): 491-494.

Bokonon-ganta A.H. and Neuenschwander P. (1995). Impact of the Biological Control Agent *Gyranusoidea tebygi* Noyes (Hymenoptera: Encyrtidae) on the Mango Mealybug, *Rastrococcus invadens* Williams (Homoptera: Pseudococcidae), in Benin. *Biocontrol Science and Technology*, 5(1): 95-108.

Chahal, B.S. & Singh, D. (1977). Bionomics and Control of Mango Shoot Borer, *Chlumetia transversa* Walk.(Noctuidae: Lepidoptera). *Indian Journal of Horticulture*, 34(2): 188-192.

Choudhary J.S., Prabhakar C.S., Sudarshan Maurya, Ritesh Kumar, Bikash Das, Shivendra Kumar, (2012). New report of Hirsutella sp. infecting mango hopper Idioscopus clypealis from Chotanagpur Plateau, India. *Phytoparasitica*, 40(3): 243-245.

Daneel, M.S., and S. Dreyer. (1998). "Biological control of the mango scale, *Aulacaspis tubercularis*, in South Africa." *South African Mango Growers Association Yearbook* 18: 52-55.

Dey, K. and Pandey, Y.D. (1987b). Evaluation of certain non-insecticidal methods of reducing infestation of the mango nut weevil, *Cryptorrhynchus gravis* (F.) in India, *Tropical Pest Management*, 33: 27-28.

George, S.A. and Abdurahman, U.C. (1985). Some aspects of the reproductive biology of Goniozus sp. (Hymenoptera: Bethylidae) on external parasite of the mango leaf webber, Lamida moncusalis Walker. Proc.Natl. Sym.Entomoph.Ins., Calicut. pp.110-15.

Grover P. (1986). Integrated control of midge pests. *Cecidol Int* 7: 1-28.

Gundappa, Shukla, P.K., Rajkumar, B., Verma, S. & Misra, A.K. (2014). Incidence of shoot gall psylla, *Apsylla cistellata* Buckton on mango in India. In *World Mango Conference, Islamia University of Bahawalpur, Pakistan.* pp. 72.

Hansen, J.D. (1993). Dynamics and control of the mango seed weevil, *Acta Horticulturae*, 341: 415-420.

Harris, K.M. & Schreiner, I.H. (1992). A new species of gall midge (Diptera cicidomyiidae) attaking mango foliage in Guam, with observations in its pest status and biology. *Bulletin of Entomological Research*. 82: 41-48.

Haseeb M., Abbas S.R. & Srivastava R.P. (2000). Missing links in the biology of mango leaf webber, *Orthaga euadrusalis* Walker (Lepidoptera: *Pyralidae*). *Insect Environment* 6: 74.

Haseeb, M. & Srivastava, R.P. (2003). Field evaluation of entomogenous fungus, *Beauveria bassiana*(Bals.) Vuill. against mango mealy bug, *Drosicha mangiferae* Green. *Journal of Ecophysiology and Occupational Health*, 3(3): 253-258.

Haseeb, M., Shukla, R.P., Misra, D. & Ram, R.A. (2003). Evaluation of biopesticides against mango mealy bug, *Drosicha mangiferae* (Green). In *Abst. Nat. Symp. Organic Farming in Horticulture*, p. 165.

Ishaq, M., Usman, M., Asif, M. & Khan, I.A. (2004). Integrated pest management of mango against mealybug and fruit fly. *International Journal of Agriculture & Biology*, 6: 452-454.

Karar, H. (2010b). *Bio-ecology and management of mango mealybug (Drosicha mangiferae green) in mango orchards of Punjab, Pakistan* (Doctoral dissertation, University of Agriculture, Faisalabad).

Karar, H., Arif, M. J., Ali, A., Hameed, A., Abbas, G. & Abbas, Q. (2012). Assessment of Yield Losses and Impact of Morphological Markers of Various Mango (*Mangiferae indica*) Genotypes on Mango Mealybug (*Drosicha mangiferae* Green)(Homoptera: *Margarodidae*). *Pakistan Journal of Zoology*, 44(6): 1643-1651.

Karar, H., Arif, M.J., Sayyed, H.A., Ashfaq, M. & Khan, M.A. (2010a). Comparative efficacy of new and old insecticides for the control of mango mealybug

(*Drosicha mangiferae*) in mango orchards. *International Journal of Agriculture and Biology*, 12(3): 443-446.

Kavitha, K., Lakshmi K.V. & Anitha V. (2005). Mango leaf webber *Orthaga euadrusalis* Walker (Pyralidae: *Lepidoptera*) in Andhra Pradesh. *Insect Environment* 11: 39-40.

Kumar D., Roy C.S., Khan Z.R., Yazdani S.S., Hameed S.F., Mahmood M. (1983). An entomogenous fungi Isaria tax parasitizing mango hopper Idioscopus clypealis. *Science and Culture*, 49(8): 253-254.

Kumar, A. Verma, T.D. & Gupta, D. (2007a). Biological Studies on mango Shoot gall psylla, *Apsylla cistellata* Buckton in Himachal Pradesh. *Pest Management in Horticultural Ecosystems*, 13(1): 13-19.

Kumar, A., Pandey, S.K. & Kumar, R. (2009). Population dynamics of mango mealy bug, *Drosicha mangiferae* Green from Jhansi, Uttar Pradesh. In *Biological Forum*, 1(2): 66-68. Satya Prakashan.

Mani, M., Krishnamoorthy, A. & Pattar, G.L. (1995). Biological control of the mango mealybug, *Rastrococcus iceryoides* (Green) (Homoptera: Pseudococcidae). *Pest Management in Horticultural Ecosystems*, 1(1):15-20.

Monobrullah, M.D., Singh, P.P. & Singh, R. (1998). Life-history and morphology of different stages of mango shoot gall psyllid, *Apsylla cistellata* Buckton (Homoptera: Pysllidae). *Journal of Entomological Research* 22: 319-323.

Patel R.K. Patel S.R. and Shah A.H. (1990). Population behaviour (sex-ratio) of mango hopper Amritodusatkinsoni (Leth.) (Jassidae: Homoptera) and their parasitism under prevailing temperature and humidity under field conditions in South Gujarat. *Indian Journal of Entomology,* 52(3): 393-396.

Peña, J.E. (2002). Integrated pest management and monitoring techniques for mango pests. In *VII International Mango Symposium* 645: 151-161.

Pena, J.E., Mohyuddin, A.I. & Wysoki, M. (1998). A review of the pest management situation in mango agroecosystems. *Phytoparasitica*, 26(2): 129-148.

Peng R.K. and Christian K. (2004) The weaver ant, *Oecophylla smaragdina* (Hymenoptera: *Formicidae*), an effective biological control agent of the red-banded thrips, *Selenothrips rubrocinctus* (Thysanoptera: *Thripidae*) in mango crops in the Northern Territory of Australia, *International Journal of Pest Management*, 50(2): 107-114.

Peng, R. and Christian, K. (2007). The effect of the weaver ant, *Oecophylla smaragdina* (Hymenoptera: *Formicidae*) on mango seed weevil *Sternochetus mangiferae* (Coleoptera: *Curculionidae*) in mango orchards in northern territory of Australia, *International Journal of Pest Management,* 53(1): 15-24.

Rafeeq, A.M. & Ranjini, K.R. (2011). Life tables of the mango leaf webber pest, *Orthaga exvinacea* Hampson (Lepidoptera: *Pyralidae*) on different host plant leaves. *Journal of Experimental Zoology*, 14: 425-429.

Rajkumar, B, Gundappa, Khan R.M. & Kumar, H.K. (2013). Integrated pest management for enhancing quality production of subtropical fruits under high density planting with canopy modification. In: canopy management and high density planting in subtropical fruit crops, 2013 Eds. Singh VK and Ravishankar H, CISH, Lucknow, 269 pp.

Sandhu, G.S., Batra, R.C., Sohi, A.S. & Bhalla, J.S. (1980). Comparison of different bands for the control of mango mealy-bug, *Drosicha mangifera* Green (Margarodidae: *Homoptera*). *Journal of Research,* Punjab Agricultural University, 17(3): 286-290.

Singh, C. (2002). Bioecology and Management of mango leaf webber *Orthaga euadrusatis* Walker (Pyralidae: Lepidoptera). Ph.D. Thesis, Govind Ballabh Pant University of Agriculture and Technology, Pantnagar, India.

Singh, G. & Misra P.N. (1978). The mango shoot gall psyllid *Apsylla cistellata* Buckton and its control. *Pesticides* 12: 15-16.

Singh, G. (2000). Physiology of shoot gall formation and its relationship with juvenility and flowering in mango. In *VI International Symposium on Mango* 509: 803-810.

Singh, G. (2003). Mango shoot gall: its causal organism and control measures. Indian Council of Agricultural Research, New Delhi, India.

Singh, G., Kumar A. & Everrett T.R. (1975). Biological observations and control of *Apsylla cistellata* Buckton (Psyllidae: *Homoptera*). *Indian Journal of Entomology* 37: 46-50.

Singh, S. (1993). Biological Control of Insect Pests. In: Advances in Horticulture: Fruit crops (Part 3) Eds. K.L. Chadha and O.P.Pareek. Malhotra Publishing House, New Delhi, India.

Schoeman, A.S. (1987). Observations on the biology of the mango weevil, *Sternochetus mangiferae*(F.), *South African Mango Growers Association Yearbook*, 7: 9.

Shukla, R.P., Tandon, P.L. and Singh, S.J. (1984). Baculovirus- a new pathogen of mango nut weevil, *Sternochetus mangiferae* (Fabricius) (Coleoptera: Curculionidae), *Current Science,* 53(11): 593-59.

Srivastava R.P. (1997). Mango insect pest management International Book Distributing Co, Lucknow.

Srivastava R.P., Tandon P.L. (1980). New records of parasites and predators of important insect pests of mango. *Entomon* 5(3): 243–244.

Srivastava R.P., Tandon P.L. (1986). Natural occurrence of two entomogenous fungi pathogenic to mango hopper, *Idioscopus clypealis* Leth. *Indian Journal of Plant Pathology*, 4(2): 121-123.

Srivastava, R.P. (1980). Efficacy of Alkathene bands to prevent ascent of mango mealy bug nymphs on mango trees. *Indian Journal of Entomology*, 42(1): 122-129.

Srivastava, R.P. (1995). Annual Report, Central Institute of Horticulture for Northern Plains, LucknowSrivastava, R.P *et al.* (1993). In: *Neem Env.*, 1: 527-534, Oxford Publisher, New Delhi.

Srivastava, R.P. and Butani, D.K. (1972). Mango hopper menace. *Entomologist Newsletter*, 2(2): 10-11.

Tandon, P.L. & Lal, B. (1978). The mango coccid, *Rastrococcus iceryoides* Green (Homoptera: Coccidae) and its natural enemies. *Current Science*, 47(13): 467-468.

Tandon, P.L., & Srivastava, R.P. (1980). New records of parasites and predators of important insect pests of mango. *Entomon*, 5(3): 243-244.

Verghese A. (1998). Management of mango leaf webber. A vital package for panicle emergence, *Insect Environment*, 4: 7.

Verghese, A., Nagaraju, D.K., Kmala Jayanthi, P.D. and Gopalakrishnan, C. (2002). Report of entomopathogenic fungus *Beauveria bassiana* (Balsamo) Vuillemin on mango seed weevil, *Insect Environment*, 8(4): 146-147.

Voute, A.D. (1935). *Cryptorhynchus gravis* F. Und die ursachen Seiner Massen Vermehrung in Java. *Arch. Neerl. Zool*, 2: 112.

Yadav J.L., Singh S.P., Kumar R. (2004). The population dynamics of the mango mealy bug (*Drosicha mangiferae* G.) in mango. *Progress Agric* 4(1): 35-37.

Vegetable Crops

10

Biointensive Integrated Pest Management of Okra

Davendra Kumar and Umesh Das

Introduction

Okra (*Abelmoschus esculentus* L. Moench) also known as lady's finger or bhendi, belongs to family Malvaceae. Tender fruits are used as vegetables and thickening of gravies and soups, because of its high mucilage content. The roots and stems of okra are used for cleaning cane juice (Chauhan, 1972). In okra several pest may damage the tender fruit that leads to reduction in yield which can be controlled by holistic approach to develop sustainable agriculture. There are several ways among this the Biointensive integrated pest management (BIPM) is one of the best way to discourage the development of pest populations and keep pesticides free so that it makes economically justified to farmers, reduce risks to human health and the environment. Biointensive integrated pest management (BIPM) is essentially a component of integrated pest management which helps to reduce the dependence on chemical pesticides thereby preventing ecological deterioration. The major elements of this approach includes many factors *viz.*, host plant resistance, use of beneficial organism, agronomic practices, biopesticides, parasitoids, predators, plant exudates etc.

Advantages of Biointensive Integrated Pest Management

1. Reduces the need for pesticides by using several Biointensive methods

2. Reduces or eliminates issues related to pesticide residue

3. Maintains or increases the cost-effectiveness of a pest management program

4. Promotes sustainable bio-based pest management

5. Protects non-target species through reduced impact of pest management activities

Table 1. Major insect pests of okra

Insect-Pest	Scientific Name	Order
Shoot and fruit borer	*Earias vittella* (Boisduval)	Lepidoptera
Fruit borer	*Helicoverpa armigera* Hubner	Lepidoptera
Jassids/Leaf hopper	*Amrascabigutulla biguttula* Ishida	Hemiptera
Red spider mites	*Tetranychus cinnabarinus* (Boisduval)	Acari
Whitefly	*Bemisia tabaci* Ishida	Hemiptera
Okra aphid	*Aphis gossypii* Glover	Hemiptera
Solenopsis mealybug	*Phenacoccus solenopsis* Tinsley	Hemiptera

Methods of Biointensive Integrated Pest Management

The most recent biointensive integrated strategy for management of pest includes several components like physical methods which include hot water treatment of planting material, soil solarization and bio-rational chemicals like pheromones. The cultural methods includes crop rotation, summer ploughing, fallowing the land, intercropping, pruning, mulching, spacing, planting date, trap cropping, use of resistant cultivars, etc. Biological methods includes use of bio-control agents (predators, parasitoids and mycorrhizal fungi), botanicals based including bio-fumigation, oil cakes, FYM, crop residues, green manuring and other organic amendments.

A. Cultural methods

a) Host plant resistance

Keeping in mind the diversity and intensity of pests in particular place, selection of resistant/less susceptible varieties/hybrids/genotypes holds well in pest management. Being completely safe, host plant resistance fits well with all other components. Pests of sucking nature can be combated to a greater extent with the adaption of resistant varieties/genotypes. Similarly, some genotypes have also been identified for the management of diseases in okra crops are as follows (Table 2).

Table 2. Tolerant genotypes of okra crops against some major insect pests

Insects-pest	Tolerant genotypes
Jassid (*Amrasca biguttula biguttula*)	IC-7194, IC-13999, New Selection, Punjab Padmini
Shoot and fruit borer (*Earias vittella*)	AE 57, PMS 8, Parkins Long Green, PKX 9275, Karnual Special

b) Planting time

Careful consideration of sowing/planting date in okra reduces the attack of fruit borer and okra shoot and fruit borer. Sowing of okra during second week of June retains less population of borers thereby enhancing yield. Thus, synchronization of most susceptible stage of the crop with the inactive period of insect pest reduces the infestation and chemical intervention.

c) Barriers

Use of nylon net as a barrier reduce the pest incidence however, the cost of nylon net is high and studies are, therefore, being conducted on the use of live barriers like maize. Presently, this technology is being popular in many parts of the West Bengal to prevent fruit and shoot borer of okra. Similarly, for the management of *Yellow vein mosaic virus* (YVMV) disease in Bhendi, growing of maize as barrier is successful to reduce the disease incidence.

d) Intercropping

Intercropping of crops with diverse plant geometry and insect pests breaks the standard mono-cropping and limits the infestation from the pest. Diverse nature of plant not only obstructs the adults from egg laying but also the release of volatile allelo-chemicals from a particular crop deters the adult insect from damaging the crops. All such planting combination enhances the activity of predators and parasites, too.

e) Trap crops

Growing of African marigold along the borders and irrigation bunds during tight bud stage functions as good trap crop to attract the adults of *H. armigera* besides it also attracts the adults of leaf miner for egg laying on the leaves.

B) Physical methods

a) Pheromones for monitoring and management of pests

Pheromone is a substance that is released into the environment by a member of a species that produces a specific response in members of the same species. For example sex pheromone (Septa) are insect specific produced artificially by the females in a laboratories which attracts the males of the same species over a longer distance and aggregation pheromone which is produced by male that attract both males, females, adults and larvae of same species over a shorter distances.

b) Insect growth regulator (IGR)

Insect growth regulators (IGRs) are pesticides that don't usually kill insects but instead affect the ability of insects to grow and mature normally. They can prevent reproduction, egg-hatch, and moulting from one stage to the next. Insect Growth Regulators (IGR) is mainly three type's viz. juvenile hormones, precocenes and chitin synthesis inhibitors. Some example of IGR which is available in market are Methoprene (fleas and beetles), Diflubenzuron (For caterpillars, beetles and flies), Lufeneron (Program®-fleas). Lufenuron produced morphological and biological abnormalities in populations of *Helicoverpa armigera* during experiments conducted by Butter *et al.* (2003).

c) Tillage

Summer ploughing is an effective practice to spoil the soil inhibiting stages of insect. Deep ploughing of the field after the harvest reduces the activity of flies and lepidopteran insects as these insects remain in the soil in earthen cocoon to complete the dormant stage of their life cycle. Similarly, summer ploughing is effective to reduce insect population because of solarization effect.

C) Biological methods

Biological control is the beneficial action of parasites, pathogens, and predators in managing pests and their damage. Like other organisms, insects also suffer from diseases caused by bacteria, viruses, fungi and nematodes.

a) Entomopathogens or Parasitoids

Okra crop control through microbial intervention is so far limited to few pests only. *Bacillus thuringiensis* (*Bt*) @ 300-500 gm is the most extensively used biocontrol agent against *E. vittella* and *Helicoverpa armigera*. Application of *Helicoverpa armigera* nuclear polyhedrosis virus (*Ha*NPV) and *Spodoptera litura* nuclear polyhedrosis virus (*Sl*NPV) @ 250-300 larval equivalents (LE) in the evening hour with some UV protectants like teepol (0.1%) and adjuvents like molasses (1%) reduces the population of the pests to a great extent. Use of entomopathogenic fungi has great potential and gaining importance against both chewing and sucking insect pests in vegetable crops. Among these *H. armigera* inokra and *Beauveria bassiana* @ 1.6×10^4 conidia/ml against whitefly and jassids on okra have been found highly effective.

b) Predator

Predators are mainly free-living species that directly consume a large number of preys during their whole lifetime. For example, *Chrysoperla zastrowisillemi*

@ 50,000 first instar larvae per hectare is an effective predator for control of white fly, aphids, jassids and eggs for some lepidopterous borers.

D) Botanical methods

a) Use of Allelochemicals

Allelochemicals are the secondary metabolites produced by organisms such as plants or microorganisms which are not needed for primary metabolism and this chemicals released from donor plants into the environment and affect growth and development of receiver pests. For example Marigold plant (*Tagetes* spp.) secret alpha-terthienyl from the root of *Tagetes* spp. that repels the several nematodes such as root-knot nematodes (*Meloidogyne* spp.) and *Pratylenchus* spp. and *Chrysanthemum cinerarifolium* produce a chemical called Pyrethrum which is used for reduction of whitefly and aphid.

b) Use of Botanical insecticides

Botanical insecticides are naturally occurring chemicals (insect toxins) extracted or derived from plants that have naturally occurring defensive properties. They are also called natural insecticides (Table 3).

Table 3. List of Botanical insecticides, sources and mode of action

Botanical insecticides	Sources	Mode of action
Pyrethrins	*Chrysanthemum cinerariaefolium*	Disrupting the sodium and potassium ion exchange process
Neem	*Azadirachta indica*	Feeding deterrent
Rotenone	*Derris* species	Inhibitor of cellular respiration
Sabadilla	*Schoenocaulon officinale*	Affect neurotransmitter action
Ryania	*Ryania speciosa*	Slow-acting stomach poison
Nicotine	*Nictiana tabacum*	Affects bonding to acetylcholine receptors
d-Limonene and Linalool	Citrus fruit peels.	Paralysis

Control of major pest of okra using Biointensive Integrated Pest Management

The primary goal of Biointensive IPM is to provide guidelines and options for the effective management of pests and beneficial organisms in an ecological context. There are several methods which can be used for Biointensive IPM are mechanical, cultural, botanical, and biological methods.

1. Okra shoot and fruit borer: *Earias vittella*

Damaging symptoms: Initial stage caterpillars bore into tender shoots and tunnel downwards. Affected shoots wilted and drooped down. During reproductive stage, they bore the fruits and feed inside it. The infested fruits become unsuitable for consumption and marketing.

Wilting of shoot   Infested fruits

Control

a) Mechanical methods

- Periodical clipping and destruction of the early infested shoots with the help of sharp knife and the damage shoot has to remove. After removal, they should be destroyed by burning or buried into the deep layer of the soil or crushing the larva at the early stage of the infestation when symptoms first appear

- Hand picking and destruction of various insect stages, affected plant parts and rosetted flowers.

- Removal and destruction of alternate weed hosts of *Earias vittella* like *Chrozophore rottlari.*

b) Cultural methods

- Deep summer ploughing by using bullock drawn or tractor mounted plough so that deep layer of soil should be exposed to kill larvae and pupae of the fruit borer to sunlight and predatory birds.

- Grow okra shoot and fruit borer tolerant genotypes such as AE 57, PMS 8, Parkins Long Green, PKX 9275, Karnual Special etc.

- Adopt proper crop rotation

- Adopt proper spacing, irrigation and fertilizer management. Avoid application of excess nitrogenous fertilizers because succulent leaves invite the infestation of *Earias vittella.*

c) Botanical methods

Use neem extract @ 1500 ppm, if pest occurrence is very severe.

d) Biological methods

- Use of *Bacillus thuringiensis* var *kurstaki* @ 500 g a.i./ha at 10 days interval showed effective control on okra shoot and fruit borer.

- Inundative release of egg parasitoid, *Trichogramma* spp. @ 2,50,000/ha 3 times at 15 days interval.

e) Behavioural methods

After 15 days of okra planting, install sex pheromone trap (Funnel trap/sticky trap/water pan trap etc.) @ 100/ha (10 m X 10 m) just above the crop canopy.

2. Fruit borer: *Helicoverpa armigera*

Damaging symptoms: In okra its infestation appears from flowering stage. The neonate larvae initially feed on the foliage and later bore into the yellowish green fruits. Damaged fruits are unfit for consumption and also invite secondary infection by other organisms lead to rotting. During the feeding, larvae characteristically thrust its head into the fruit leaving the rest of its body outside.

Caterpillar of Helicoverpa

Control

a) Mechanical methods

- Removal and destruction of crop residues after harvest to avoid the carry over population of *Helicoverpa armigera* to next season.

- Removal of terminal shoot of marigold (Pinching) at 30-40 days after planting to reduce *Helicoverpa* oviposition and also to encourage sympodial branching which help to bears more number of flower.

- On colonization of pest population on trap crop hand picking and destruction of larvae from trap crop are recommended.

- Removal and destruction of alternate weed hosts of *Helicoverpa armigera* like *Abutilon indicum, Chrozophore rottlari and Solanum nigrum.*

b) Cultural methods

- Summer deep ploughing to expose soil inhabiting / resting stages of insects and nematode population.

- Use of trap crops like okra, marigold (*Tagetes* spp.), early pigeon pea, coriander, maize crops along the borders is recommended. Insects feeding on these crops must be removed and destroyed.

- During planting, sow one row of 40 days old seedling of marigold every after 16 days of 25 days old seedling of okra.

- Adopt proper crop rotation

- Adopt proper spacing, irrigation and fertilizer management. Avoid application of excess nitrogenous fertilizers.

c) Botanical methods

- Spray neem seed kernel extract (NSKE) 5% as a strong oviposition deterrent or neem oil 3% starting from one month after planting at 15 days interval.

- If the infestation is severe, spray the botanicals insecticide such as Azadirectin @ 1500 ppm at weekly intervals.

d) Biological methods

- Application of *Helicoverpa armigera* nuclear polyhedrosis virus (HaNPV) @ 250-500 ml/ha depending upon the crop growth with jaggery and teepol in evening hours at 7th and 12th week after sowing.

- ULV spray of NPV at 3×10^{12} POB /ha with 10% cotton seed kernel extract (CSKE), 10% crude sugar, 0.1% each of Tinopal and Teepol.

- Inundative release of egg parasitoid, *Trichogramma* spp., at 6.25 cc/ha at 15 days interval 3 times from 45 DAS.

- Inundative release of egg-larval parasitoid, *Chelonus blackburnii* and predator, *Chrysoperla carnea* at 100000 / ha at 6th, 13th and 14th weeks after sowing.

- *Trichogramma brassiliensis*, 2.5 cc/ha. once in 10 days (Egg parasitoid).

e) Behavioural methods

Set up pheromone traps @ 12/ha at one month after transplanting for monitoring *Helicoverpa armigera* at a distance of 50 m. When the trapped moths are 8/day necessary action may be taken. Trapped moths should be removed daily.

3. Jassids/Leaf hopper:*Amrasca bigutulla biguttula*

Damaging symptoms: Damage caused by both the nymphs and adults by sucking the sap from the underside of the leaves. Affected leaves initially turned yellow and then converted to pinkish red and become brittle and crumple followed by upward curling of leaves. Severely affected plants have stunted growth.

Damage of jassid

a) Mechanical methods

- Collection and destruction of leaves infested with jassids

- Removal and destruction of nearby weed which acts as alternate hosts of jassids from the fields, neighbouring areas and maintaining field sanitation.

- Use silver or grey coloured mulches of 100 micron or 400 gauges that can repel jassids.

b) Cultural methods

- Grow jassids tolerant genotypes of okra such as IC-7194, IC-13999, New Selection, Punjab Padmini etc.

- Keep plants well irrigated and avoid excessive applications of nitrogen fertilizer which may promote higher populations of jassids

- Old and spent plant can harbour jassids, so their removal and disposal is sometimes recommended

- Avoid close planting in jassids prone areas or during summer seasons

c) Botanical methods

Spray Azadirachtin 0.5% is very useful to control jassid.

d) Biological methods

*   *Chrysoperla zastrowisillemi* is an effective predator for control of jassid when the first instar larvae are released @50,000 /ha.

*   *Beauveria bassiana* @ 1.6 x 10⁴ conidia/ml against jassids have been found highly effective.

4. Red spider mites: *Tetranychus cinnabarinus*

Damaging symptoms: Mites normally inhabit the under surface of leaves and spin the webs. The plants often remain covered with dense webs under which both the developing stages and adults damage the okra crops by sucking the cell sap. Besides, dust particles get adhered to webs and retard normal physiological activities of the plant leading to reduction in plant growth and yield. The affected leaves turn yellow, dry up and finally wither away giving an unhealthy appearance.

a) Mechanical methods

*   Flooding the growing area to provide the proper moisture to soil surface.

*   Removal and destruction of crop residues after harvest to avoid the carry over population of mites to next season.

b) Cultural methods

*   Growing areas should not be dry provide adequate water especially during of drought period, spider mites will often be worse on drought stressed plants first.

*   Inside greenhouses maintain higher humidity levels that can reduce spider mite populations and damage.

c) Botanical methods

Spray neem seed kernel extract (NSKE) @5% or neem oil @3%.

d) Biological methods

*   The number one predator for two spotted spider mites is the persimilis mite (*Phytoseiulus persimilis*). This mite works best at temperatures between 55°F to 85°F. The recommended release rate is 1 mite per square foot for effective controlling.

*   During hot summer months, the swirski mite (*Amblyseius swirskii*) works best on two spotted spider mites. Swirskii works best at temperatures above

68°F. They are very aggressive and can be applied generally at a rate of 5-10 mites per square foot.

- Application of *Beauveria bassiana* (1.0%) affects the young stage of mites.
- Application of *Bacillus thuringiensis* subsp. *kustaki* (0.3-0.4%) affects all stage of mites.

e) Behavioural methods

Set up *Bacillus thuringiensis* mite trap @ 12/ha for monitoring mite at a distance of 40-60 mt is beneficial.

5. Okra aphids:*Aphis gossypii*

Damaging symptoms: Direct feeding through sucking the plant sap by both nymph and adult causes stunted growth of the plants and poor flower bud formation. Infested seedlings lose their vitality and distorted. Beside these, they also excrete honey dew which favours the development of sooty mould on the foliage there by reducing the photosynthetic activity of the plants.

a) Mechanical methods

- Collection and destruction of aphid infested twigs.
- Remove the little leaf affected plants.
- Clipping the tips of the seedlings up to 2 inches prior to flowering to remove the egg masses of aphids if any.

b) Cultural methods

- Growing castor along the borders and irrigation bunds as trap crop for aphids.
- Deep summer ploughing on bright sunny days during the months of May or June should be done to expose soil inhabiting or resting stages of insects and nematode population. The field should be kept exposed to sunlight for at least 2-3 weeks.

c) Botanical methods

- Spray fish oil resin soap 25 kg / ha @1 kg in 40 litre of water.
- Spray neem oil 3% plus teepol (1 ml/litre) or spray neem seed kernel extract 5 %.
- Spray NSKE (5%) or neem oil (3%) alternatively.
- Apply neem cake @ 250 kg/ha as a basal dosage.

d) Biological methods

- *Chrysoperla* spp. (Green lace wing) 5000-10000 eggs /ha after starting of first sign of aphid presence 3–4 times in 15 days.
- Application of *Verticillium lecanii* (0.5-1.0%) affects the all growth stages of aphid.
- Conserve the predator's *viz.,* spiders, coccinellids and wasps to check the population of aphids.

e) Behavioural methods

- Use of light trap @ 12/ha to monitor and trap the aphids.
- Set up the yellow sticky traps @ 25/ha to monitor the activity of pest and to synchronise the botanical pesticide application, if need be, at the maximum activity stage.

6. White fly

Damaging symptoms: Both the nymphs and adults suck the sap and also secrets the honeydew on which black sooty moulds develops that reduce the photosynthesis of the plants.

Adult whitefly   Damage symptoms (YVM)

a) Mechanical methods

- Removal and destruction of alternate weed hosts of white fly like *Abutilon indicum*, *Chrozophore rottlari*, *Solanu nigrum* and *Hibiscus ficulensus* from the fields and neighbouring areas and maintaining field sanitation.

- Use aluminum foil or reflective plastic or silver coloured mulches that can repel adult whiteflies.

b) Cultural methods

- Collection and destruction of leaves infested with white fly.

- Keep target areas free of weeds that can serve as whitefly hosts.

- Avoid excess use of nitrogen fertilizer, including manures as succulent growth will increase whitefly population.

c) Botanical methods

- Spray NSKE 5% or neem oil (5 ml/litre) or 5 % notchi leaf extract or 5% *Catharanthus rosea* extract.

- Spray fish oil rosin soap 25 g/ litre and add teepol as wetting agent.

- If the infestation is severe, spray the botanicals insecticide such as Azadirectin @ 1500 ppm at weekly intervals.

d) Biological methods

- *Chrysoperla* spp. (Green lace wing) 5000 - 10000 eggs /ha, 3-4 times in 15 days after first sign of whitefly presence.

- *Delphastus pusillus* is the most whitefly-specific predator. Recommended release rates are 7-10 per m^2.

- *Eretmocerus eremicus* is the most effective parasitoid. Recommended release rates in the greenhouse are three wasps per m^2 every 1-2 weeks after starting of first sign of whitefly presence.

e) Behavioural methods

- Monitoring the activities of the adult white flies by setting up yellow pan traps and sticky traps @ 12/ha at 1 foot height above the plant canopy.

- Locally available empty yellow palmoline tins coated with grease / Vaseline/ castor oil on outer surface may also be used to catch white flies.

References

Alavo, T.B. (2006). Biological control agents and environmentally-friendly compounds for the integrated management of *Helicoverpa armigera* Hübner (Lepidoptera: Noctuidae) on cotton: Perspectives for pyrethroid resistance management in West Africa. *Archives of Phytopathology and Plant Protection* 39(02): 105-111.

Butter, N.S., Singh, G. and Dhawan, A.K. (2003).Laboratory evaluation of the insect growth regulator lufenuron against *Helicoverpa armigera* on cotton. *Phytoparasitica*, 31(2): 200-203.

Campos, E.V., Proença, P.L., Oliveira, J.L., Bakshi, M., Abhilash, P.C. and Fraceto, L.F. (2019). Use of botanical insecticides for sustainable agriculture: future perspectives. *Ecological Indicators*, 105: 483-495.

Chadha, K.L. (2019). *Handbook of Horticulture in Vol 2.* New Delhi: Indian Council Agricultural Research.

El-Wakeil, N.E. (2013). Retracted article: Botanical Pesticides and Their Mode of Action. *GesundePflanzen*, 65(4): 125-149.

Mohan, S., Devasenapathy, P., Vennila, S. and Gill, M.S. (2015). Pest and disease management: organic ecosystem. Coimbatore: Tamil Nadu Agricultural University.

Mouden, S., Sarmiento, K. F., Klinkhamer, P. G., and Leiss, K. A. (2017). Integrated pest management in western flower thrips: past, present and future. *Pest management science*, 73(5): 813-822.

Nivsarkar, M., Cherian, B. and Padh, H. (2001). Alpha-terthienyl-A plant derived new generation insecticide. *Current Science,* Bangalore, 81(6): 667-672.

Rai, A. B., Loganathan, M., Halder, J., Venkataravanappa, V., and Naik, P. S. (2014). Eco-friendly approaches for sustainable management of vegetable pests. *Technical Bulletin*, 53 pp.

Srinivasan, K., Krishnamoorthy, P.N. and Prasad, R. (1993). Evaluation of different trap crops for the management of fruit borer, H. *armigera* on tomato. In Abstract, Golden Jubilee Symposium. Horticultural Research: Changing Scenario Bangalore, India. 259 pp.

11

Biointensive Integrated Pest Management of Brinjal

Ponnusamy N., Anil Kumar and Abbas Ahmad

Introduction

Brinjal is an important vegetable crop grown extensively throughout India. In India, brinjal is cultivated in around 730.4 ha with a production of 17.5 million tonnes in the year 2017-18. It is also called as poor man's vegetable. It is consumed in variety of ways depending upon the eating habits of the different parts of the country. It is grown extensively in Bihar, Orissa, Karnataka, Tamil Nadu, Andhra Pradesh, Maharashtra, West Bengal, Uttar Pradesh and states with matching climatic conditions in the tropics and subtropics.

Brinjal is an important vegetable crop, but insect pests are making more losses in brinjal cultivation. In brinjal, more than 36 pests have been reported from the time of planting to harvest (Regupathy *et al.*, 1997). Among all insect pests shoot and fruit borer (*Leucinodes orbanalis*), hadda beetle (*Epilachna vigintioctopunctata*), jassid (*Cestius phycitis*), Ash weevil (*Myllocerus subfasciatus),* white fly *(Bemisia tabaci),* lace wing bug *(Urentius hystricellus)* are important (Barker and Pritchrad, 1960).

Despite serious nature of the pest, its management tactics by and large is limited to frequent sprays of chemical insecticide without much impact on the yield. Such practice of pesticide usage is detrimental to environment, also increases the cost of production and chances of insecticide residues in the fruit. About 47 per cent insecticides are used for management of fruit and shoot borer out of the total pesticide consumed in vegetables. In this background major emphasis, now-a-days, is being given on Biointensive integrated pest management as alternate to the insecticides for management of any pest. More than a dozen parasitoids and three pathogens have been recorded as natural enemies of insect pest. But the extent of parasitisation/predation under field condition is very low. In this context, inundative release of bio-agents, particularly egg parasitoids in pest management may be more advantageous.

Important insect pests of brinjal

S.No.	Common name	Scientific name	Family	Order
1.	Shoot and fruit borer	*Leucinodes orbanalis*	Pyralidae	Lepidoptera
2.	*Stem borer*	*Euzophera perticella*	Pyralidae	Lepidoptera
3.	*Hadda beetle*	*Epilachna vigintioctopunctata*	Coccinellidae	Coleoptera
4.	*Ash weevil*	*Myllocerus discolor*	Curculionidae	Coleoptera
5.	*Lace wing bug*	*Urentius hystricellus*	Tingidae	Hemiptera
6.	*White fly*	*Bemisia tabaci*	Aleyrodidae	Hemiptera
7.	Leaf hopper	Cestius phycitis	Cicadellidae	Hemipttera
8.	*Red spider mite*	*Tetranychus urticae*	*Tetranychidae*	*Acari*

1. Shoot and fruit

Damage and symptoms

- Brinjal shoot and fruit borer is the most destructive pest of brinjal.

- Its starts infestation after transplanting of the seedlings to till harvesting of fruits.

- Newly hatched larva bore into the growing tips of young shoots during earlier stage of plants (Hegde *et al.*, 2009).

- During young stage of plant, drooping of shoot is the typical damage symptom of this pest and the affected shoots wither and die away.

- At the time of flowering and fruiting stage, the larvae prefer flower buds and young fruits and bore into the young fruits.

Larvae of shoot and fruit borer   Damage symptom

- Firstly, the larva makes a very small hole around the calyx and goes inside the fruit.

- Since, it completes its larval stage within the fruit and the last instars matured larva comes out from the fruit for pupation.

Life cycle

- Female lays about 250 eggs within 2 to 5 days of their emergence.
- The eggs are laid singly on ventral surface of leaves, shoots and flower buds
- They also lay occasionally on fruits and calyx during March-April.
- Mostly eggs are laid in the morning hours.
- The eggs are flat and white in colour.
- The early instars larvae hatches out from eggs within 3-5 days.
- The larvae enter the plant tissues immediately after their hatching.
- After five moultings it becomes fully matured.
- The fully developed larva is stout, pink colored with brown head.
- The larval body bears wart all over the body through which hairs protrudes out.
- The total period of larvae about 10-15 days lies inside of the fruit.
- The matured larva comes out of the host tissues and forms pupa among the fallen leaves or on the surface of stem and fruits of the host plant.
- Pupation occurs inside a grey tough cocoon.
- The total pupal period is about 6-8 days.
- The adult moths live for 2-5 days.
- The total life cycle of shoot and fruit borer is about 21-43 days.

Biointensive Integrated Management

- The infested leaves, shoots or fruits should be plucked and destroyed.
- The fallen fruits and dropping leaves and shoots containing larvae should be removed.
- In heavy infestation, the whole infested plant should be uprooted and destroyed.
- Avoid monocropping pattern of brinjal and ratooning.
- Grow resistance varieties.
- Use light traps @ 1/ha to attract and kill the moths.
- Release egg parasitoids *Trichogramma chilonis* @1.0 lakh/ha.
- Setting up pheromone lures i.e Lucinlure

- Spray *Bt* formulations of *B. thuringiensis* var. *kurstaki* such as Dipel @ 1.5 to 2 ml /L of water.
- Spray insecticide starting from one month after planting at 15 days interval with Azadirachtin 1.0% @ 1.0-1.5 l/ha

2. Stem Borer

Damage and Symptoms

- Larva starts boring into the stem at ground level.
- Mostly they bore in the branching area or in leaf axils, and seal the entry holes with excretory materials.
- Larvae feed used to downward along the length of the main stem, which results in stunted growth, or wilting and withering of the whole plant.

Life Cycle

- Adult female lay eggs singly or in groups on the tender leaves, shoots and petioles.
- The cream colored eggs are elongate and flat.
- The egg period varies from 3-9 days.
- The larva is white or yellowish white in colour and also several bristly hairs present on body.
- The larval period is about 29-47 days.
- Larvae pupate within silken cocoon inside the feeding tunnel.
- The total pupal period is around 7-14 days, whereas adult longevity is about 4-8 days for male and 5-13 days for female.

Biointensive Integrated Management

- Remove and promptly destroy the infested plants.
- Avoid ratoon cropping.
- Light trap @1/ha to attract and kill adults.
- Protect the population of parasitoids such as *Pristomerus euzopherae*. Reduced use of synthetic pesticides may enhance the activities of these natural enemies.
- Apply Neem cake in the soil to reduce the incidence of stem borer.
- Spray any one of the insecticide starting from one month after planting at 15 days interval with Neem oil 1.0 L, NSKE 5%, Azadirachtin 1.0% 1.0-1.5 L with 500-750 L water/ha

3. Hadda beetle

Damage and Symptoms

Epilachna vigintioctopunctata

- Both adult and grubs scrap the lower epidermis of leaves in characteristic manner leaving behind stripes of uneaten areas.

- The leaves give a stifled appearance. In severe infestation all leaves may be eaten off leaving only the veins intact (Skeletonization) and plants may wither.

Life Cycle

- The female starts egg laying in the month of March-April.

- A gravid female laid 200-370 eggs on an average in 6-7 batches during her life span (Tara and Sonia Sharma, 2017).

- The eggs are laid in cluster of 45-50 in numbers on the lower surface of the leaves.

- The eggs are cigar shaped yellowish in colour and are arranged side to side on the surface of the leave in erect position.

- The larvae emerge from eggs in 3-4 days in summer and in 4-9 days in winter. Its total larval period would be around 15 days.

- The grubs are oval, fleshy and yellow in colour bearing multi branched spines on the body surface.

- The larval period lasts for 9-18 days during which it passes through four different stars.

- The larva changes into pupa. The pupation takes place on leaf surface or on stem or at the base of the plants.

- At time of pupation the larva attaches its last abdominal segment to the surface of host plant by means of sticky secretion. Pupa is oval in shape and dark in colour.

- The total pupal period would be 3-4 days but in some cases it may extend further.

- Total life period is around 17-18 days in summer but in winter it may prolong up to 50 days.

Biointensive Integrated Management

- Collect and destroy adult beetles, grubs and pupae.

- Thorough irrigation of infested crop can minimize the increase in pest population.

- Remove the alternate host plants.

- Introduction of *Tetrastichus ovulorum* Ferr. and *Achrysocharis appannai* to the crop to parasitize the eggs of hadda beetle. The grubs are parasitized by *Solindenia vermai, Pleurotropis epilachinae, Tetrastichus* sp. while pupa is parasitized by *Pleurotropis foveolatus.*

4. Ash Weevil

Damage and Symptoms

- Damage symptoms include leaves notched in the margin and defoliated leaving only the midrib, accompanied by presence of sandy brown weevils mottled with grey and white on the foliage.

- The infested leaves are like half-shaped moon in the margin of leaves.

- The subterranean grubs cause damage to the roots which results in wilting, drying and death of the infested plants (Shanmugam *et al.*, 2018).

Life Cycle

- The fecundity range of adult female is from 58 to 109 eggs. They lay eggs in soil and plant debris.

- Embryonic development ranged from 9 to 11 days (Drekiæ *et al.*, 2014)

- After hatching of grubs from egg, feed on roots.

- The total larval period is about 30-45 days.

- Pupation takes place in soil in earthern cocoons. Adult weevil life span is around 10-12 days.

Biointensive Integrated Management

- Collect and destroy adult weevils.

- Apply Neem cake @ 500 kg/ha at the time of last ploughing

- Entomopathogenic nematodes *Heterorhabitis* and *Steinernema* are potential biological control agents against insect pests (Nagesh *et al.*, 2016).

5. Lace wing bug

Damage and Symptoms

- Both nymphs and adults are usually found on the ventral surface of leaves.

- Infested leaves are leaves with yellow patches, soiled with exuviae and excreta; yellowish brown, flat, lace-like adults suck the sap from dorsal surface and spiny black nymphs on the under surface of leaves.

- It can destroy up to 50 per cent in severe case.

Life Cycle

- A female bug lays about 75 shining white nipple shaped eggs singly in the tissues on the underside of leaved.

- Incubation period of *U. hystricellus* of 5-7 days at a temperature of 26.6 to 40 °C and the nymphs feed gregariously on the lower surface of the leaves, but fully developed nymphs are found feeding and moving about individually on the lower surface as well as on the upper surface of leaves (Padwal *et al.*, 2016).

- Nymphal period is about 10-23 days and full grown nymphs yellowish brown and are stoutly built, with prominent spines.

- Adults are measures about 3 mm in length and are pale yellow colored on the dorsal side and on the ventral side.

Management

- Plant grown under partial shade experience less damage by lace bugs than when they are grown in locations more exposed to direct sunlight and higher temperatures.

- Provide adequate irrigation

- Natural enemies of lace bugs include parasitic wasps, predatory assassin bugs, lacewing larvae, lady beetles, jumping spiders, pirate bugs, and mites.

- Apply ultra-fine horticultural oil or insecticidal soap to control damaging populations of this pest. Be sure to direct the spray to the upper and lower leaf surfaces.

- Weekly sprays of an aqueous neem seed extract (50 g/l) and 2% neem oil gave good results.

6. Whitefly

Damage and Symptom

White fly

- Hot and dry climate favor the multiplication of whitefly.

- It is active during the day and settles on lower leaf surfaces at night.

- Nymphs and adults suck the plant sap and reduce the vigor of the plant.

- In severe infestations, the leaves turn yellow and drop off. When the populations are high

- They secrete large quantities of honeydew, which favors the growth of sooty mould on leaf surfaces and reduces the photosynthetic efficiency of the plants.

Life cycle

- The females mostly lay eggs near the veins on the underside of leaves. Each female can lay about 300 eggs in its lifetime. Eggs are small (about 0.25 mm), pear-shaped, and vertically attached to the leaf surface through a pedicel.

- Egg period is about three to five days during summer and 5 to 33 days in winter.

- After hatching nymphs moves on leaf surface to find a suitable feeding site.

- The nymphs are flattened, oval-shaped, and greenish-yellow in color. Nymphal period is about 9 to 14 days during summer and 17 to 73 days in winter.

- Adults emerge from puparia through a T-shaped slit, leaving behind empty pupal cases or exuviae.

- The whitefly adult is a soft-bodied, moth-like fly. The wings are covered with powdery wax and the body is light yellow in color. The wings are held over the body like a tent. The adult males are slightly smaller in size than the females.

- Adults live from one to three weeks.

Biointensive Integrated Management

- The field selected for brinjal or seedling production should be clean and not be located near any host plants and weeds.

- Grow eggplant seedlings in insect-proof (50-64 mesh) net houses, net tunnels, greenhouses, or plastic houses.

- Use yellow sticky traps at the rate of 1-2 traps/50-100 m² to trap the whiteflies.
- Plant fast-growing crops like maize, sorghum, or pearl millet in the border of the field to act as barriers to reduce whitefly infestations.
- Spray Neem seed kernel extract @ 5%.

7. Leaf hopper

Damage symptom

Leaf hopper

- Both nymph and adult suck sap from plant leaves and tender parts.
- Upward curled and scorched leaf due to hopper damage.

Life cycle

- Female laids eggs along the midrib and lateral veins of the leaves.
- The egg period varied from 4 -11 days.
- The nymphs similar to the adults stage, but lack wings.
- The nymphal period varies from one to four weeks depending on the climatic condition.
- The adults may live for 1-2 months.

Biointensive Integrated Management

- Select tolerant or resistant cultivars
- Destruction of crop residues
- Crop rotation with Gramineae and crucifers Establish sunflower as a barrier crop around the field two weeks before planting Plant okra around the field
- Use recommended fertilizer and manage the crop well to increase plant vigour
- Mixed cropping with Gramineae and crucifers
- Conserve natural vegetation to encourage natural enemies like Lace wings, Spiders, Lady bird beetle etc
- Water spray to reduce the population buildup
- Use yellow stick traps and water trays to reduce pest population
- Spray neem seed kernel extract @30-40 g/ l
- Spray Neem seed kernel extract @ 5%.

8. Red Spider mite

Damage symptom

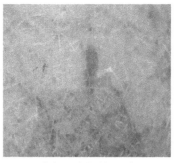

Adult Red Spider mite

- Both nymphs and adult cause damage by sucking the cell sap from undersurface of leaves. Affected leaves become mottled, turn brown and fall.

- Different stages of mites are found in colonies covered by white-silky webs on lower surface of leaves.

- Low relative humidity favors mite multiplication.

- The mite very active from March – October.

Life cycle

- Female lays 60-80 eggs per female.

- Eggs are spherical and hatch in 2-6 days.

- The duration of larval period is 6-8 days.

Control

- Collection and burning of severely infested plant parts reduces further multiplication of mites.

- Proper irrigation and clean cultivation are essential to keep the pest population under control.

- Spray Neem seed kernel extract @ 5%.

References

Baker, E.W. and Pritchrad, A.E. (1960). The tetranychoid mites of Africa. *Hilgardia*, 29(11): 455-574.

CABI. Crop protection compendium. CAB International (Available at: http://www.cabicompendium.org/cpc Retrieved on March 15, (2012), 2007.

Drekiæ, M., Poljakoviæ pajnik, L., Vasiæ, V., Pap, P., and Pilipoviæ, A. (2014). Contribution to the study of biology of ash weevil (Stereonychus fraxini De Geer). *Šumarski list*, 138(7-8): 387-395.

Halder, J., Kushwaha, D., Dey, D., Tiwari, S.K., Rai, A.B., and Singh, B. (2017). Biology of stem borer, Euzophera perticella (Lepidoptera: Pyralidae) and association of endoparasitoid Pristomerus euzopherae (Hymenoptera: Ichneumonidae) in grafted and ratoon brinjal crop. *Indian Journal of Agricultural Sciences*, 87(6): 801-3.

Hegde, J.N.; Girish, R. and Chakravarthy, A.K. (2009). Integated management of brinjal shoot and fruit borer, *Leucinodes orbonalis* (Guenee). In: Proceeding International Conference on "*Horticulture for Livelihood Security and Economic Growth*", November 9-12, 2009, University of Agricultural Science, Bangalore pp: 1103-1107.

Mabberley, D.J. (2008). *Mabberley's Plant-Book*. Cambridge University Press.

Nagesh N., Krishnakumar N.K., Shylesha A.N., Srinivas N., Saleem Javeed and Thippisamy (2016). Comparative virulence of strains of entomopathogenic nematodes for the management of eggplant grey weevil *Myllocerus subfasciatus* Guerin (Coleopetra; Curculionidae). *Indian Journal of Experimental Biology,* 54: 834-842

Onekutu, A., Omoloye, A.A., and Odebiyi, J.A. (2013). Biology of the eggfruit and shoot borer (EFSB), Leucinodes orbonalis Guenee (Crambidae) on the garden egg, Solanum gilo Raddi. *Journal of Entomology*, 10(3): 159-162.

Padwal, K.G., Sharma, S.K., and Singh, S.K. (2016).Major insect pests of brinjal and their management.*Popular Kheti*, 4(2): 48-53.

Patel, R.C. and Kulkamy, H.L. (1955). Bionomics of *Urentius echinus* Dist. (Hemiptera: Heteroptera : Tingidae) an important pest of brinjai *(Solanum melongena* L.) in North Gujarat. *J. Bombay Natural History Society,* 53(1): 86-96.

Shankar, U., Kumar, D. and Gupta, S. (2010). Integrated pest management in brinjal. Technical Bulletin No. 4. pp 16. Sher-e-Kashmir University of Agricultural Sciences and Technology of Jammu.

Shanmugam, P.S., Indhumathi, K., and Sangeetha, M. (2018). Management of ash weevil *Myllocerus subfasciatus* Guerin-Meneville (Coleoptera; Curculionidae) in Brinjal. *Journal of Entomology and Zoology Studies*, 6(6): 1230-1234

Tara, J.S., and Sharma, S.O.N.I.A. (2017). Biology and life cycle of *Henosepilachna vigitioctopunctata* Fabricius, a serious defoliator of bitter gourd (*Momordica charantia*) in Jammu region (Jammu and Kashmir) India. *Indian Journal of Science Research*, 13(1): 199-203.

12

Biointensive Integrated Pest Management of Cucurbitious Crops

Ayan Das and Sudarshan Chakraborti

Introduction

Cucurbits belong to the family cucurbitaceae; constitute the largest group of summer vegetables grown all over the world which includes about 118 genera and 825 species. Cucumbers, muskmelons, watermelons, squashes, gourds, luffagourd, tissel-gourd and pumpkins are commonly grown cucurbits are mostly important one. These crops are attacked by a variety of insect pest and diseases throughout its growth period; generate wounds that help to transmit of viral particles, bacteria and invasion of fungal pathogens. Major insect pests include cucumber beetles, red pumpkin beetles, fruit flies, epilachna beetles, squash bugs, aphids, white flies, squash vine borers, two-spotted spider mites, and nematodes affects significant damages from seeding stage upto harvest. Insect pest infestations in cucurbits bring about heavy losses through reduction in yield, lowered quality of produce, defoliation of leaves, damage roots or flowers, contribute to poor crop standard increased cost of production and harvesting besides expenditure incurred on materials and equipments to apply control measures.

Fruit flies

Bactrocera cucurbitae(Coquillet) and *B. ciliates* (Loew) *B. zonata*

(Saunders) *Bactrocera dorsalis* (Hendel) (Tephritidae: *Diptera*).

Identification of pest: The adult, which is noticeably larger than a common house fly, has a body length of approximately 8.0 mm in length ; the wing is about 7.3 mm and is mostly hyaline.

Fruit fly (*Bactrocera cucurbitae*)

The variable color of the fly is found, but there are prominent yellow and dark brown to black markings on the thorax.

Distribution: Fruit fly is distributed widely in temperate, tropical, and subtropical regions of the world (Sapkota *et al.*, 2010).

Host range: 81 host plants and is a major pest of cucurbitaceous vegetables viz. melons, bittergourd, muskmelon, snap melon, snake gourd, cucumber, tissel gourd, mango, guava, etc.

Loss: The extents of loss vary between 30% and 100%, depending upon the cucurbit species and the season (Shooker *et al.*, 2006).

ETL: 5% fruit damage.

Status of the pest: Major pest of almost all cucurbits (Kapoor, 2002).

Bioecology

The female fly lays singly or in groups of 4-10 eggs into the tender fruit that are whitish elongate, cylindrical in shape, slightly curved and tapering at both ends. Eggs are 1.0-1.5 mm long. Full grown maggots measured 5-10 mm long, cylindrical in shape, tapering anteriorly, blunt at posterior end and pale-white in colour. Pupae are 5-8 mm long, barrel shaped and brown to coriaceous in colour. The fully fed maggots come out of the fallen fruits and pupate 10-15 cm soil depth. Where the fruits do not fall, the maggots pupate inside the fruits (which are not common) or come out, drop down and pupate in the soil. Pre-oviposition, egg, maggot and pupal periods last for 9-21, 1-1½, 3-9 and 6-8 days respectively. During winter the larval and pupal stages are extended up to 3 and 4 weeks, respectively. A single life-cycle is completed in 10 - 18 days but it takes 12-13 weeks to complete a single life-cycle in winter. Adult longevity vary between 2-5 months; females live longer than males. The nature the population is generally low during dry weather and increases rapidly with adequate rainfall. Adult with hyaline wings with costa band broad and prominent, anal stripes well developed and hind cross veins thickly margined with brown and grey spots at the apex and face with two black spot.

Symptoms of damage

*   Females lay their egg and causing damage in fruits.
*   Oozing of resinous fluid from fruits at infection.
*   Maggots feed on the pulp of the fruits.
*   Sometimes pseudo-punctures (punctures without eggs) have also been reported on fruit skin, which reduces the market value of the produce.

- Oozing of resinous fluid from fruits.
- Distorted and malformed fruits.
- Premature dropping of fruits.
- The attacked fruits decay because of secondary bacterial infection.
- After the first shower of the monsoon, the infestation often reaches 100 per cent.

Management

- Field sanitation will be adopted at the community level; it will significantly reduce the melon fly population, having a better reduction of pest load in crop field than conventional one. Burying infested fruit 15-45 cm deep in the soil reduces the survival rate of melon fly maggots and prevents adult fly emergence (Klungness *et al.*, 2005; Kumar *et al.*, 2011).
- Expose the pupae by ploughing and turning over soil after harvest helps in exposing the pupae for desiccation and predation by predatory birds.
- Cover the fruits as far as possible at the earliest with paper cover or perforated polythene tubes of required size.
- Apply neem cake at 100 g per pit about 10 days before sowing so as to manage the pupae present in soil.
- Apply *Beauveria bassiana* and *Paecilomyces lilacinus* in the basins as soil application.
- Suspend different types of traps at 2.5 meter height in field.

Banana trap- Take 100 ml of water and add 20 g ripened banana pulp. Mix 10 g melted jaggery and poison it with 2 drops of insecticide -malathion. Transfer it to a medium sized (500 ml) bottle provided with 4 holes in the middle, large enough for the fly to enter. Suspend the traps in trellis.

Fish meal trap- Place 10 grams of moistened dried fish, preferably powdered, in a coconut shell, add 2 drops of malathion and suspend in trellies. Trap adult male fruit flies using cue-lure (pheromone) plywood blocks containing 6:4:1 mixture of ethyl alcohol, cue-lure and malathion respectively. Reset traps at 4 months interval. Hang this pheromone traps @ 10 numbers per/ha.

Water bottle bait Installed with cue-lure (as MAT) saturated wood blocks (ethanol/ cue- lure/carbaryl in a ratio 8:1:2) at 25 traps/ha prior to flower initiation is for trapping male flies.

- Use of a repellent (NSKE 4%) could enhance trapping and luring in bait spots.

- Spray *Beauveria bassiana or Verticillium lecanii* @ 20 g or 5 ml/Liter. (Konstantopoulou and Mazomenos 2005).

- Resistance germplasm against melon fruit fly, *Bactrocera cucurbitae* are Kerala collection1, Faizabad collection 17 for bittergourd; IHR 83, 86, Arka Suryamukhi for pumpkin; Pusa Smooth Purple Long for bottle gourd, ArkaTinda for round melon.

Pumpkin beetle

Red pumpkin beetles: *Aulacophora foveicollis* (Lucas),*Raphidopalpa foveicollis* (Lucas); *Aulacophora cincta* Fabr. (Grey) and *A. lewisii* (Baly) *(=A. intermedia* (Jacoby) (blue beetle); Purple bettele: *A. cincta*, Ash beetle : *A. intermedia* (Chrysomelidae: Coleoptera).

Red pumpkin beetle

Identification of pest: Red pumpkin beetle adults are oblong and 5-8 mm long. Their dorsal body surface is brilliant orange red ventral surface is black, which is covered with soft whitish hairs.

Distribution: Pumpkin beetle is distributed widely in temperate, tropical, and subtropical regions of the world. *R. foveicollis* is found in almost all states of India though it is more abundant in northern states in association with *A. foveicollis, A. lewisii.*

Host range: Pumpkin, ash gourd bottle gourd, cucumber, muskmelon, watermelon, tinda, ghia tori, beans and other cucurbits.

Loss: 30%–100% in the field (Khan *et al.*, 2012).

ETL: 5 insects per sweep or 5 holes per leaf.

Bioecology: Female lay yellowish pink spherical eggs are in the soil which turns orange after two days and a beetle may lay 150-300 eggs. The egg period varies from 5-8 days. The grubs become full grown in 13-25 days and pupate inside the soil. The pre-pupal period is 2-5 days. The pupal period ranges from 7-17 days. Total life cycle occupies takes 32-65 days to complete. In a year there may be 5-8 generations of the insect. They are active from March to October and the peak period of activity in April to June.

Pest status: A common and regular pest of a wide range of cucurbitaceous vegetables.

Nature of damage

- The grubs as well as adult beetles scrap and feed voraciously on the green matter of the leaf and skeletonize it in a characteristic manner leaving the upper epidermal tissue intact and causes irregular holes on leaves and also feed on floral parts (Rath *et al.*, 2002; Mohasin and De, 1994).

- Grubs develop in the soil and feed on underground parts, causing wilting of plants.

- Sometimes grubs enter the fruits touching the ground.

- The infested portion start rotting due to saprophytic fungus infection at the injury site and the fruits become unfit for consumption.

- Whereas the adults feed on initial stages of crop growth and attack is so severe that the entire crop requires re-sowing.

Management

- In the initial stages, it is good to collect the beetles and destroy them.

- Preventive measures like burning of old plants, plowing, and harrowing of field after harvest, proper crop rotation with non-cucurbitaceous crop are followed for the destruction of adult, larvae, and pupae. (Khan, 2012).

- Use of neem oil cake in the soil is effective in killing the pest larvae.

- Spray *Beauveria bassiana* (2×10^6 cfu/g) against red pumpkin beetle.

- The application of plant extract of *Parthenium sp.* was found to be highly effective in controlling the red pumpkin beetle (Ali *et al.*, 2011).

- Plants products like leaf extract of *Ageratum conyzoides* @ 0.06% and oil of basil @ 1.5 ml/liter of water acts as repellent.

- Use of egg parasitoid: *Trichogramma* spp., larval Parasitoids: *Brachymeria tachardiae, Trichospilus pupivora.*

- Neem extracts mixed with benzene acts as repellency against beetles (Khan and Wasim, 2001).

- Use of resistant varieties: Kerala collection-1, Faizabad-1 (Bitter gourd), NB-29 (Bottle gourd), No. 21, 19, 32, 40 and 47 (Muskmelon) Selection-1, Candy.

Epilachna (Hadda) beetle

Henosepilachna vigintioctopuctata (Fabricius) (Coccinellidae: *Coleoptera*)

Identification of pest: Adults are medium in size (8-10 mm), which bears 12-28 black spots on their back. The full grown grubs (6-7 mm) are yellow, hump shaped and spiny in appearance. The yellow colored pupa is 6-7 mm in length and hemispherical in shape.

Epilachna grub

Distribution: Australia, Africa, East Indies, America, India and Japan (Rajgopal and Trivedi 1989). It is an important pest of solanaceous and cucurbitaceous crops in mid-hills and plains of India (Kumar and Kumar 1998).

Status of the pest: Major insect pests in cucurbit.

Host: Bitter Gourd, Cucumber, Squash, Watermelon, Pumpkin, Brinjal, Potato, Tomato.

Loss: The reduction percentage plant product varies 10-25%.

ETL: Average population including grubs and pupae 2 per leaf.

Bioecology

The female lays 300-400 eggs in clusters on the undersurface of the leaves that hatch into yellowish larvae. The larva (7-9 mm long) is found on the underside of leaves and is yellow with branched black spines covering the body and each about 2 mm high and 1 mm diameter. Full-grown larvae pupate below the leaf or at the base of the stems. The pupa hangs from the leaf, is yellow in color, and lacks spines. The development stages are completed within 4-6 weeks under optimal conditions. Adult beetles are orange in colour with 28 black spots on their back. Adults overwinter under loose tree bark or under leaf litter near the edge of fields.

Nature of Damage and Symptoms

- Both grub and adult stages cause damage by scrapping and feeding voraciously green matter leaving parallel bands of uneaten portion in between from leaves which dry and die up (Rath *et al.*, 2002; Mohasin and De, 1994).

- Adults make semicircular cuts in rows in leaf portion and also feed on flowers parts.

- The affected leaves become lace-like in appearance, turn brown, dry up and drop prematurely (Ghosh and Senapati, 2001).

- Sever infestation take in early stages can kill the plants, whereas the older vines show stunted growth and poor yield growth and poor yield.

Management

- Harrowing and destroying vines and larvae after harvesting early cucurbits can help to minimize pest population.

- Crop rotation to distant fields tends to limit colonization and population build up (Boucher, 2014).

- Row covers can protect cucurbit crops from the beetles.

- Spores of *B. thuringiensis* were sprayed and gave effective mortality of epilachna beetle. (Venkataraman *et al.,* 1962).

- Seed extracts of *Annona squamosa* (3 mL/L of water) as botanical pesticide help in reducing population buildup to the extent of 76% (Mondal and Ghatak, 2009).

- Leaf extracts of three plants, namely*, Ricinus communis*, *Calotropis procera*, and *Datura sp* significant toxicity against by adversely affecting both oviposition and egg hatching besides prolonged larval duration, pupae formation, and adult emergence (Islam *et al.*, 2011).

- *NeemAzal* (5 mL/L of water) and petroleum ether extracts of rhizome of *A. calamus* (2 mL/L of water), *Tephrosia* leaf extract (20g/100 mL water) effectively controls *Epilachna* beetle by killing adults and inhibiting pupae formation (Rahaman *et al.*, 2008).

Cucumber Beetle

Spotted Cucumber Beetle: *Diabrotica undecimpunctata howardi* Barber. Striped Cucumber Beetle: *Acalymma vittatum* F. (Chrysomelidae: *Coleoptera*)

Identification: The larva of spotted cucumber beetle (*D. undecimpunctata*) is yellowish white with a brown head and is 0.6–1.2 cm long with a brown spot at the tail end. The adult beetle has a black head with black antennae and 12 black spots on its yellow-green body.

Striped cucumber beetle   Spotted cucumber beetle

The larva of *A. vittatum* is white and 8.5 mm in length, while adults are about 5 mm long with three longitudinal black stripes on its top wings.

Distribution: Spotted and striped cucumber beetles are native insects distributed throughout the tropical to subtropical countries including India (Capinera, 2008).

Host range: Cucurbitaceae, groundnut soybean *Phaseolus vulgaris* and other legumes, maize sweet potato etc.

Pest status: Serious pest of cucurbit.

Loss: Attacks on older plants result in a general leaf parching. 100% loss of watermelon seedlings as a result of attack by *D. u. howardi* in Canada (Beirne, 1971).

Bioecology

Spotted beetle females eggs typically hatch within 6–9 days (Webb 2010, Alston and Worwood, 2008). Eggs are generally oval, 0.7 x 0.5 mm, light yellow when first oviposit, but become darker yellow as they age. The surface of the egg is covered with tiny hexagonal pits. Larvae have a yellowish white, wrinkled body, 12-19 mm long, with six very small legs, and a greyish-brown head. Length about 6.3 mm, width about 3 mm. In *D. u. undecimpunctata*, the abdomen is greenish-yellow with 11 black spots on the elytra while in *D. u. howardi,* it is yellow to yellowish-red with 12 large black spots. The head, antennae and legs are entirely black (*howardi*) or with some greenish-yellow (*undecimpunctata*). *D. undecimpunctata* is a larger and heavier beetle than *D. virgifera*. The striped beetle female deposit their eggs at the base of host plants, below the ground surface. Upon hatching (8-10 days) the larvae migrate to the root system and feed upon roots for 2-6 weeks, during which time they may consume the entire root system. Larvae pupate in the soil, emerging as adults in about one week. The adult beetles are 1/4" long and yellow-green with black longitudinal stripes.

Symptoms of damage

- The larvae feed on roots and tunnel through stems (Sorensen, 1999) also feed on emerging seedlings below the soil surface; as a result, the plant is either killed or may have stunted growth.

- Adult beetles feeding on cotyledons and young leaves can cause reduction in crop stand and delay growth.

- Adults feed on fruits and cause scars on the fruit surface, thereby lowering marketable quality of fruits.

- They transmit diseases like bacterial wilt (*Erwinia tracheiphila*) squash mosaic virus (Diver, 2008).

- The bacterium causes bacterial wilt (*E. tracheiphila*) that overwinters only in the intestines of some of the striped cucumber beetles between 1% and 10% (Mitchell and Hanks, 2009).

Management

- Early plowing-disking, delayed planting and heavy seed rates help to minimize the impact of these beetles (Sorensen, 1999).

- Application of Organic manure reduces its infestation (Yardim *et al.*, 2006).

- Companion plants such as radish, aluminum plastic mulch increased cantaloupe yield and vine cover and reduced cucumber beetle populations.

- Straw mulches were also effective at managing beetles in many ways: slowing down beetle movement, providing refuge for predators (wolf spiders).

- Growers should select cucurbit varieties with lower cucrbitacin levels to decrease their attractiveness for cucumber beetles.

- Indole, cinnamaldehyde alone or in combination with trimethoxybenzene are attractive kairomones and have been used under field conditions to attract beetles (Capinera, 2008).

- Kaolin clay, pyrethrum, neem product are some of the organic chemicals that can be used to manage cucumber beetles (Synder, 2012).

- Natural enemies viz. tachinid flies, *Celatoria diabrotica* and fungi (*Beauveria*), and nematodes can be applied for manage the beetle population (Capinera 2008; Cline 2008).

- Conservation of predatory spider (*Hognahelluo* and *Rabidosarabida*) has been shown to feed heavily on these beetles in cucurbit crops (Snyder and Wise, 2001).

Snake gourd semilooper (*Anadevidia peponis* F.) Plusia peponis (Noctuidae: *Lepidoptera*)

Identification: It is a specific pest of snake gourd and the larvae defoliate the plant considerably, if infestation is high. The brown moth has shiny brown forewings.

Distribution: It is found in south-east Asia, including Japan, Taiwan, Australia and widely

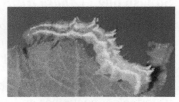

Anadevonia peponis

distributed in India.

Host range: The larvae feed on cucurbitaceae species, including watermelon, cucumber, pumpkin, bittergourd, cho-cho, snake gourd etc.

Pest status: Regular pest of snake gourd all over the country also infest other cucurbits.

Bioecology

Female moths lay white spherical eggs singly on undersurface of tender leaves. Semilooper is greenish with white longitudinal lines and black tubercles with thin hairs on it. Last abdominal segments are humped measures 35-40 mm long. Egg, larval, pupal periods last for 3-6, 9-14 and 5-13 days respectively. Larva elongate bright green with a pair of thin white longitudinal lines on the dorsal side. Pupation takes place inside the leaf fold. Pupa is greenish but it turns dark brown before emergence of adult. Adult is a brown moth with shiny brown forewings. Adult lives for 3 -7 days and females lays upto 366 eggs in its lifetime.

Symptoms of damage

* The caterpillar cuts the edges of leaf lamina, folds it over the leaf and feeds from within leaf roll.

* It is a specific pest of snake gourd and the larvae defoliate the plants considerably if infestation is serious.

* The early sown cucurbits are so severely damaged that they have to be resown.

Management

* Collect and destroy the caterpillars.

* Encourage activity of *Apanteles taragamae, A. plusiae.*

* The larvae of Snake gourd semilooper were found parasitized by three different species of parasitoids, belonging to Braconidae, Bethylidae and Eulophidae and all were larval parasitoids.

Squash bug : (*Anasa tristis, A. armigera* G.) (Coreidae: *Hemiptera*)

Identification: The adult *Anasa tristis* is a greyish-brown, somewhat flattened insect reaching a length of about 1.5 cm and a width of 0.75 cm. There is often a row of alternate brown and gold spots along the margin of the abdomen.

Host range: Cucurbitaceae, but most often occurs on pumpkins and squashes, watermelon, cucumber and cantaloupe melon etc.

Pest status: Occasional pest in cucurbits.

Bioecology: The adult female lays about eighteen eggs. The eggs are oval, somewhat flattened and bronze in colour, and are deposited on the underside of the leaves of the host plant. They may be clustered close together or more widely dispersed but are often regularly arranged. The eggs hatch after seven to nine days into nymphs which have five instar stages. The first instar nymphs are green and about 2.5 mm in length. Each successive instar is larger and less hairy and grey. The fifth instar is grey, with developing wing pads and about 10 mm in length. The complete nymphal stage lasts about 33 days.

Symptoms of damage

- *Anasa tristis* feeds by sucking sap from leaves, fruits.
- Insects inject toxic saliva into the plant tissues which causes them to wilt, darken in colour and die the wilting of plants is sometimes called "anasa wilt."
- Sometimes one plant or part of a plant can be heavily attacked while surrounding plants are untouched.
- Besides the direct damage their feeding causes to the plant, these insects can act as vectors for cucurbit yellow vine disease (Bruton *et al.*, 2003).

Management

- Remove old cucurbit plants after harvest.
- Keep the garden free from rubbish and debris that can provide overwintering sites for squash bugs.
- During the growing season, pick off and destroy egg masses.
- Use protective covers such as row covers in gardens where squash bugs have been a problem in the past.
- Using a trellis for vining types of squash and melons can make them less vulnerable to squash bug infestation.
- Some squash varieties, including Butternut, Royal Acorn, and Sweet Cheese, are more resistant to squash bugs.
- The parasitic tachinid fly Trichopoda pennipes, which lays its eggs on squash bug attacks nymphs and adults, has been introduced into California and may be found in some gardens. (Worthley,1923).

Snake gourd stem weevil: (*Baris trichosanthis* S.) (Curculionidae: *Coleoptera*)

It is a small weevil that feeds upon green matter of foliage. Adult female lays egg in nodes, the incubation period is 5-6 days. The grub bore into the stem or the petiole for about weeks and causes withering of leaves. They pupate in the bore hole itself and emerge in about a week. The adult is black in colour.

Squash Vine Borer or Clear Winged Moth : (Melittia cucurbitae and Melittia eurytion) (Sesiidae: Lepidoptera)

Identification: Adult moths are stout, dark gray moths (1.25 cm long) with orange abdomen, having black dots, hairy red hind legs, opaque front wings, and clear hind wings with dark veins.

Host range: It is a native pest of cucurbits, especially summer squash, pumpkins, and gourds. Cucumbers and melons are less affected.

Pest status: Minor and occasional pest.

Bioecology: The small, brown, flat eggs about 1 mm in size are laid individually on leaf stalks and vines. They hatch in 7–10 days and the newly born larva immediately bores into the stem. The cream-colored larvae, about 2.5 cm in size, enter and feed in the stems of cucurbit vines blocking the flow of water to the rest of the plant, which cause wilting of plants and finally mortality (Britton, 1919). A larva feeds for 14–30 days before exiting the stem to pupate in the soil (Delahut, 2005).

Symptoms of damage

*   Symptoms appear in midsummer when an entire plant wilts suddenly.
*   Infested vines usually die beyond the point of attack.
*   Sawdust-like frass (insect excrement) near the base of the plant is the best evidence of squash vine borer activity (McKinlay and Roderick, 1992).
*   The primary feature for distinguishing squash vine borer from Bacterial wilt and *Fusarium* wilt is frass accumulating at the entrance to the larval tunnel.

Management

*   Early planting helps to escape losses as crop matures before egg laying.
*   Floating row covers placed over the crops prevent the moths from laying eggs in the initial stages (Welty, 2009).
*   *Steinernema carpocapsae* or *Steinernema feltiae* applied to the stem and soil provided control similar to a conventional insecticide (Canhilal and Carner, 2006).
*   Application of *Bacillus thuringiens* (Bt) *is* provided similar control to that of the conventional insecticide (Canhilal and Carner, 2007).

Caterpillars

Melonworms or rindworms (Diaphania hyalinata) Cucumber moth/Pumpkin caterpillar (Diaphania indica=Cryptographis indica (Saunders) (Crambidae: Lepidoptera).

Diaphania catterpillar

Identification: *D.indica* adults have translucent whitish wings with broad dark brown borders. The body is whitish below, and brown on top of head and thorax as well as the end of the abdomen. There is a tuft of light brown "hairs" on the tip of the abdomen, vestigial in the male but well developed in the female. In *D. hyalinata* the wings appears pearly white centrally, and slightly iridescent, but are edged with a broad band of dark brown.

Distribution: This moth occurs tropical and subtropical regions, native to southern Asia, India, Fiji, Australia, Japan, Africa, Sudan, Pakistan, and South America.

Host range: Cucurbitaceae: *Cucumis* sp., *Cucurbita* sp., *Lagenaria siceraria, Luffa cylindrica, Trichosanthes anguina, T. cucumerina.* Snake cucumber was found to harbor larvae the most, while pumpkin was the least preferred (Mohaned *et al.*, 2013).

Loss : 9%–10% direct yield reduction due to fruit damage while indirect yield loss up to 23% due to foliage damage and has been reported (McSorley and Waddill, 1982).

Pest status: It is regular and significant pest of cucurbits and some other plants.

Bioecology

A female lays an average of 187.1 eggs. Eggs are oval and flattened in shape, about 0.7 mm in length and 0.6 mm in breadth. The average egg larval, pre-pupal, pupal stages lasts for 4.75, 11.9, 1.3, 9.4 days respectively. They are laid at night in a cluster of 2-6 eggs on different plant parts like buds, stems, and underside of leaves and hatch within 3-4 days (Capinera, 2005). The larva is yellow green, about 2.5 cm long, and has fine yellow stripes running down its back in the last instar. The mean developmental period from oviposition to adult emergence ranges from 23- 33 days with an average of 27.35 days. The severity observed during April-September on cucurbits.

Symptoms of damage

- The caterpillars lacerate and feed on chlorophyll content of foliage; later fold and web the leaves together and feed within.

- High populations defoliate plants giving a lace like appearance as only leaf veins remain intact.

- The larva may feed on the surface of the fruit or even burrow into the fruit in case of non- availability of foliage or non-preference crops like cantaloupe, leading to the name rindworm.

Management

- Early plantings often escape serious damage except in tropical areas where melonworms overwinters (Capinera, 2005).

- Intercropping with corn and beans helps to reduce its damage (Letourneau, 1986).

- Squash is the most preferred host among cucurbits and can be used as a trap crop (Smith, 1911).

- *B. thuringiensis* is commonly recommended for suppression. The Bt variety 'kurstaki' is most effective on caterpillars of pickleworm and melonworm (Zhender, 2011).

- Important parasites like *Apanteles* sp., trichogrammatids (all hymenopterans) also suppress the pest (Capinera 1994).

- Medina-Gaud *et al.* (1989) reported that predators such as *Calosoma* spp., *Harpalus* spp., soldier beetle, and *Solenopsis invicta* (red fire ant) caused mortality of melonworms up to 24%.

- Establishment of bird perches; common mynah and other carnivorous birds predates on larvae of *D. indica.*

Serpentine leaf miner: (*Liriomyza trifolii, Liriomyza sativae*) (Agromyzidae: Diptera)

Identification: Adult is pale yellow fly. Larva is orange yellowish and apodous. The shiny black mesonotum of *Liriomyza sativae* is used to distinguish this fly from the closely related American serpentine leafminer, *Liriomyza trifolii* which has a grayish black mesonotum.

Distribution: Tropical and subtropical area in world, widely distributed in India.

Host range: Tomato, Brinjal, Cow pea, French bean, Squash, Leafy vegetables, cucurbits.

Loss: 30 to 60% yield loss may occur when infestation is high.

Pest status: Regular pest in cucurbitaceous crop.

Bioecology

Females can produce 600 to 700 eggs over their life span. The white, elliptical eggs measure about 0.23 mm in length and 0.13 mm in width. Larvae attain a length of about 2.25 mm. Initially the larvae are nearly colorless, becoming greenish and then yellowish as they mature. The larva usually emerges from the mine, drops from the leaf, and burrows into the soil to a depth of only a few cm to form a puparium. The reddish brown puparium measures about 1.5 mm in length and 0.75 mm in width. After about nine days the adult emerges from the puparium, principally in the early morning hours, and both sexes emerge simultaneously. The adults are principally yellow and black in color.

Symptoms of damage

*   Foliage punctures caused by females during the acts of oviposition or feeding, may cause a stippled appearance on foliage.

*   The irregular mine increases in width from about 0.25 mm to about 1.5 mm as the larva matures, and is virtually identical in appearance and impact with the mines of *Liriomyza trifolii.*

*   Leaves are often with serpentine mines followed by drying and dropping of leaves due to infestation.

Management

*   Collect and destroy mined leaves.

*   Conservation of Lace wings, lady beetle, spiders, fire ants, dragonfly, robber fly, praying mantis etc. can effectively reduce the pest load.

*   Place yellow sticky traps above the crop to attract males, in particular, as they are more mobile and fly low between the plants.

*   Clear weeds especially broadleaf weeds from around crops as the weeds act as alternative hosts of the leafminer.

*   Make sure plants are not water stressed as healthy plants can better tolerate the leafminer.

*   Soil solarization can be done to destroy initial puparium stages.

*   Mulching and staking of vegetables may influence both leafminers and their natural enemies.

Thrips: *Thrips tabaci, Thrips palmi* Karny; Tobacco Thrips, *Frankliniella fusca* Hinds (Thripidae. *Thysanoptera*)

Identification: Nymphs: Yellowish, Adult: Dark coloured with fringed wings

Distribution: Worldwide distribution found throughout in temperate, tropical and subtropical zones.

Host range: Vegetables including garlic, onion, and pepper, cucurbitaceous, crucifer; many herbaceous ornamentals etc.

ETL: 10 thrips per leaf or 15-20% affected plants.

Pest status: Major and regular pests in cucurbit ecosystem.

Bioecology

Adults overwinter in trash, under bark, and in other protected places. Adults become active in the spring and lay eggs in the tissues of plants. The eggs hatch into nymphs, which begin feeding in flowers, buds, and leaves. When mature, nymphs drop to the ground and molt into adults. Under favorable conditions, a life cycle may require only 2 weeks.

Symptoms of damage

- Silvery streaks on leaf surface.

- Its feeding causes bronzing of foliage and destruction of vine tips, leading to limited canopy development.

- Feeding damage to developing leaves leads to scarring.

- Pre-mature dropping of flower.

Management

- Keep plants well irrigated, and avoid excessive applications of nitrogen fertilizer, which may promote higher populations of thrips.

- Prune and destroy injured and infested terminals when managing.

- Floating row covers (vented polyethylene, spunbonded polyester, point-bonded polypropylene) can be placed on top of beds to eradicate thrips population.

- Conservation of natural enemies , predatory thrips, green lacewings, minute pirate bugs, mites, and certain parasitic wasps help to control plant-feeding thrips.

- Azadirachtin (neem oil)-application can be beneficial for its management.

- *Beauvaria bassiana*-Some formulations can be similarly used as organic use.

- Use blue sticky traps @ 15/ha.

- Release larvae of *Chrysoperla cornea* @10,000/ha.

Aphid: Green peach aphid (*Myzus persicae* S.) and Melon aphid (*Aphis gossypii* G.) (Aphididae:*Homoptera*)

Identification: Aphids are soft-bodied insects, almost pear-shaped about 3 mm long. Most species have a pair of tube like structures called cornicles projecting rearwards from their abdomen.

Distribution: Distributed worldwide, but are most common in tropical and subtropical zones.

Aphid

Host range : *Aphid* is highly polyphagous on summer hosts, which are in over 40 different families, including Brassicaceae, Solanaceae, Poaceae, Leguminosae, Cyperaceae, Convolvulaceae, Chenopodiaceae, Compositae, Cucurbitaceae and Umbelliferae. Summer hosts include many economically important plants.

Loss: Generally 10-29% loss can be observed affected plants even upto 100 percent loss can be occurred in heavy infestation and virus transmission.

Pest status: Major and serious pests of cucurbits.

Bioecology

Adult females give birth to nymphs, which are always wingless. Aphid populations can increase rapidly in a short period of time. They tend to cluster on succulent plant parts and take just 10–12 days to complete one generation and reproduce over 20 generations annually under mild climates (Capinera 2005).

Symptoms of damage

- Aphids pierce the plant tissue and extract sap, which causes a reduction of plant growth and vigor, mottling, yellowing, browning, curling, or wilting of leaves, malformation of flower buds and fruits which result in low yields and sometimes death of plant.

- They also excrete honeydew and growth of sooty molds (fungi) on leaves and other plant parts, thereby hindering photosynthesis by blocking light.

- Aphid is considered to be the most important vector for the transmission of viruses viz. cucumber mosaic virus(CMV), watermelon mosaic virus(WMV), zucchini yellow mosaic virus(ZYMV), and papaya ringspot virus(PRSV).

- The virus infection causes mottling, yellowing, or curling of leaves and stunted plant growth and in severity plant death may occur.

Management

- Installation of yellow sticky traps in field.
- Floating row covers or reflective mulches may help exclude or repel aphids and check transmission of viruses (Stapleton and Summers, 2002; Barber check 2014).
- Conservation of aphid predators viz. lady beetles (Koch 2003), minute pirate bug, larvae of the syrphid fly (Laska *et al.*, 2006), and green (*Chrysoperla carnea*) and brown (*Hemerobius sp.*) lacewing reduces aphid population.
- Aphids are very susceptible to fungal diseases in humid weather. Some fungi that infect and provide biological control of aphids are *B. bassiana*, *M. anisopliae*, *Verticillium lecanii* (Hall, 1982) can be applied three times at an interval of 5–7 days for effective control.
- Application of Tobacco decoction with a mixture of ordinary bar soap can be very effective for controlling aphids and other soft-bodied insects infesting vegetable crops.
- Application of NSKE 5% can effectively reduce pest load?
- Neem oil+garlic emulsion (2%) can be better suited in pest management programme.

Whitefly

Tobacco whitefly (*Bemisia tabaci* G.), silverleaf whitefly (*Bemisia argentifolii* Bellows & Perring.) and greenhouse whitefly (*Trialeurodes vaporariorum*) (Aleyrodidae: *Diptera*)

Identification: The body and wings of adult flies are covered with fine whitish powdery wax.

Distribution: Since the whitefly is predominately associated with areas exhibiting tropical/subtropical climates in world and also in India.

Whitefly nymph and adult

Host range : Tomatoes, squash, poinsettia, cucumber, eggplants, okra, beans, and cotton.

Loss: 100 percent loss can be occurred if infested at initial stage and viral transmission occure. ETL : 5-10 adults or 20 nymphs per leaf during early morning.

Pest status: Major pest of cucurbitaceous crop.

Bioecology

A female can lay around 300 eggs (Nyoike, 2007). The eggs are oval and are laid into a slit in the leaf surface. The eggs are initially white, changing to brown, and are hatched within 8–10 days. The first instar, crawler is the only mobile instar having legs and antennae that moves to look for feeding sites, while the other instars are sessile and complete their life cycle on the same leaf (McAuslane and Smith 2000). Whiteflies can complete one generation in about 3–4 weeks.

Symptoms of damage

- Nymphs and adults cause chlorotic spots by sucking cell sap from leaves resulting yellowing and drying of leaves.

- Damage may be more severe on younger plants compared to plants near harvest.

- Excreted honeydew promotes the growth of sooty mold (*Capnodium* spp.) on leaves and economic plant parts cases reduction of photosynthesis.

- They also damage the plant by transmitting viruses namely, cucurbit chlorotic yellow virus, cucurbit yellow stunting disorder virus, beet pseudo yellow virus, and lettuce infectious yellows virus in field or greenhouse-grown cucurbits (Abrahamian and Abou- Jawdah, 2014).

Management

- Avoid excess irrigation and nitrogen.

- Good sanitation, removal of weeds and host crop residues in winter and spring crops is also required for the maintenance and control of the fly population.

- Silver/aluminum cover mulches can repel the adult silverleaf whitefly.

- Monitor the incidence using yellow sticky trap @ 12/ha.

- Natural enemy conservation; spider, crysoperla can helpful in maintenance of whitefly population.

- *B. bassiana* is only an effective biological control agent in conditions of low temperatures (maximum of 20 °C) and a humidity level greater than 96%.

- In the family Cucurbitaceae, vegetables such as watermelon and squash contract squash vein yellowing virus (SqVYV) by the silverleaf whitefly can be utilized as repellent crop.

- Spray Neem oil 3 ml/lit + Teepol 1 ml/lit or NSKE 5% (50g /lit).

- Application of Tobacco decoction with a mixture of ordinary bar soap can be very effective for controlling whitefly and other soft-bodied insects infesting cucurbitaceous crops.
- Application of NSKE 5% can effectively reduces pest load?
- Neem oil + garlic emulsion (2%) can be better suited in pest management programme.
- Application of Fish Oil Rosin Soap 20 g/lit can be helpful in minimizing pest load.

Mites

Two-spotted spider mite or red spider mite (*Tetranychus urticae* Koch)Arachnida: Acari are the most serious, while the other species viz. red-legged earth mite (*Halotydeus destructor*), broad mite (*Polyphagotarsonemus latus*), blue oat mite (*Penthaleus major*), and clover mite (*Bryobia cristata*) can be found in cucurbitaceous ecosystem are of less importance.

Identification: The two-spotted spider mite adults are typically pale green, orange, or yellow in color with two dark spots on the body and about 0.3–0.5 mm long. During unfavorable or cold weather conditions, they may change to an orange/red color and are commonly known as red spider mites.

Distribution: Distributed worldwide, but are most common in tropical and subtropical zones.

Host range: They have such a wide host range that they usually start feeding wherever they land.

Loss: Near about 14% reductions in the total leaf area that results in significant yield loss can be noticed (Park and Lee, 2005).

Pest status: Minor pest in cucurbit ecosystem.

Bioecology: Female lays around 200 eggs. The eggs are small, round, translucent, straw-colored, shiny, and pearl-like and are laid near veins on the undersurface of leaves. These eggs hatch into a nymph with having only three pairs of legs. The larva molts into protonymph, having four pairs of legs then molts into deutonymph and turning into an adult. Depending upon the ambient temperature and other biotic and abiotic factors, the life cycle may vary from 6 to 40 days. Under optimum temperature of more than 27°C and relative humidity less than 50%, the mites can complete their life cycle within 5–7 days. The life cycle is faster at higher temperatures, and hence, the multiplication of spider mites is alarmingly rapid in greenhouses and can complete 5–6 generations per year.

Symptoms of damage

• The initial symptoms appear as pale or bronzed areas along the midrib and veins of the leaves. Leaves turn speckled, yellowish, or grayish in appearance. These infestation causes mottling of leaves with silvery-yellow appearance and leads to premature leaf fall (Gupta 1995).

• Infestations usually begin on lower leaves of plants, then progress upwards. Lower leaves turn stippling and webbing first.

• Mite infestations usually start on the field edge and move toward the center over time.

• Loss of leaves can lead to sunburning and have a significant impact on yield.

Management

• Excessive nitrogen fertilization may cause population buildup.

• Cultural control methods for mites include weed control, crop rotation, clean fallowing, mixed cropping, proper cleanup at the end of the crop is to be done to reduce initial infestations in the next crop (Murphy *et al.*, 2014).

• Irrigation with an overhead sprinkler may provide some short-term relief of mite infestations. Good water management increases plant tolerance to these pests.

• Ten thousand predators *Phytoseiulus persimilis, Amblyseius californicus, Amblyseius fallicus* over 200 m2 of crop should be released at the first sign of TSM for effective management option.

• Minute pirate bugs (*O. tristicolor*), big-eyed bugs (*Geocoris* spp.), six-spotted thrips (*Scolothrips sexmaculatus*), western flower thrips (*Frankliniella occidentalis*),lady beetles/spider mite destroyers (*Slethorus* spp.), and lacewing larvae (*C. carnea*) also predate upon mites.

• *B. bassiana* effectively controls mites and should be applied three times at a 5–7-day interval.

• TSMs population was achieved through the four successive alternate sprays of botanical insecticide Nimbecidine and entomopathogenic fungus(Bio-Catch) Dimetry *et al.*, (2013)

• Recommended plant extract-based formulations include garlic extract, clove oil, mint oils, rosemary oil, and cinnamon oil (Godfrey, 2011).

• Neem oil provides control, when combined with a suitable surfactant and diluted with water.

Nematodes (*Meloidogyne* sp., *Pratylenchu* ssp., *Trichodoru* ssp., *Paratrichodorus* sp., *and Longidorus africanus*) **Nematoda.**

Identification: Nematodes are typically microscopic, elongated roundworms. Plant parasitic nematodes obtain their food only from living plant tissues.

Distribution: Mostly found upto 30-50 cm below soil surface even upto 1m below soil surface. Distributed throughout the world tropical ,subtropical temperate zone.

Host range: Most cucurbits, especially muskmelon, cucumber, pumpkin, bottle gourd, and bitter gourd, are extremely susceptible to plant parasitic nematode. Both *Cucumis* species (cucumber and melon) are the more preferred hosts for nematode invasion and reproduction than zucchini squash, followed by watermelon (Lopez-Gomez and Verdejo-Lucas, 2014).

Loss: Upto30% yield loss (Westerdahl and Becker, 2011) in case of RKN. Davis (2007) reported 24%–30% reduction in fruit weight of watermelon in Lesion nematode.

ETL: 1-2 larvae per of soil.

Pest status: Root-knot nematode is a problem mainly in areas with lightly textured or sandy soils.

Bioecology

Each mature female can produce more than 1000 eggs during its relatively short life. The eggs hatch immediately followed by three or more molting before they become adults. The juvenile root-knot nematodes enter the roots and continue feeding and complete most of their life cycle within the roots of their host plant although they can survive in the soil as eggs or as second stage juveniles. These are encased in a gelatinous sac that protects them from dehydration. The optimum soil temperature for root-knot nematodes development is 25°C–28°C, and they complete their life cycle in 3–4 weeks at this range. However, development will take more time at lower temperatures (Westerdahl and Becker, 2011).

Symptoms of damage

- They establish a permanent feeding site of "giant cells". The "knots" (hyperplastic galls) formed on infected roots are due to an increased number of plant cells around the feeding site.

- The symptoms include stunting, flagging, and chlorosis between veins of the leaves, which may be confused for nutrient deficiency symptoms.

- Plants show premature wilting and watering cannot relieve either the problem or symptoms.Plants have poor root system and can be uprooted easily.

Symptoms usually occur in patches of non-uniform growth rather than as an overall decline of plants within an entire field.

- Root-knot nematodes (Meloidogyne spp., RKN) are one of the most destructive pathogens of vegetables; even low nematode levels can cause high yield losses (Mukhtar *et al.*, 2013).

- **Lesion nematodes** are migratory endoparasites that invade roots. They move and feed within the root cortex.

- Infestation may cause reddish brown to dark brown lesions on roots.

- Plants stressed by inadequate nutrition or moisture may be more susceptible to damage by root-knot nematodes and are also vulnerable to infection by other pathogens (Noling, 2009).

Management

- Various cultural practices like crop rotation, summer ploughing, fallowing, sanitation, water management, organic manuring and mulching help in the management of nematodes.

- Crop rotation with non-host crops is very effective, for example, grasses, cowpea, and winter cereals are generally poor hosts and little nematode reproduction occurs during the cold winter months.

- Follow a 3-year or longer crop rotation with non-host crops. Marigold serves as a good rotation crop as its roots exude chemicals (allelopathic effects) that kill root-knot nematodes (Wang *et al.*, 2007).

- Higher soil organic matter content protects plants against nematodes by increasing soil water-holding capacity and enhancing the activity of naturally occurring biological organisms that compete with nematodes in the soil.

- Many cowpea cultivars are poor hosts to root-knot nematodes (Wang and McSorely 2004) and should be used as cover crops to protect the main crop. Resistance to root-knot nematodes was reported in cucumber cv. 'Capris' (Khelu *et al.*, 1989).

- Various fungal agents viz. *P. lilacinus, T. viride* and *Aspergillus flavus* parasitizes the eggs *M. incognita* and is effective in biomanagement of nematode population (Xalxo *et al.*,2013).

- Several microbial pathogen formulations like *Pasteuria penetrans,B. thuringiensis, Trichoderma, Hirsutella, Verticillium, Paecilomyces lilacinus* were found to be highly effective in the control of nematodes (Messenger and Braun, 2000) can be applied in nematode management.

References

Abrahamian, P.E. and Abou-Jawdah. Y. (2014). Whitefly-transmitted criniviruses of cucurbits: Current status and future prospects. *Virus Disease*. 25(1):26-38.

Ali, H., Ahmad, S., Hassan, G., Amin, A. and Naeem. M. (2011). Efficacy of different botanicals against red pumpkin beetle (*Aulacophorafoveicollis*) in bitter gourd (*Momordica charantia* L.). *Journal of Weed Science Research*. 17(1): 65–71.

Alston, D.G., Worwood, D.R. (2008). Western striped cucumber beetle, Western spotted cucumber beetle (*Acalymmatri vitatum* and *Diabrotica undecipunctata undecipunctata*). Utah Pests Fact Sheets. ENT-118-08. Utah State University Extension and Utah Plant Pest Diagnostic Laboratory. (12 Sep 2013).

Arant, F.S. (1929). Biology and control of the southern corn rootworm. *Bulletin of the Alabama Agricultural Experiment Station*, No. 230.

Barbercheck, M.E. (2014). Biology and management of aphids in organic cucurbit production systems. www. extension.org/pages/60000/biology-and-management-of-aphids-in-organic-cucurbit-production-systems.

Beirne, B.P. (1971). Pest insects of annual crop plants in Canada. I. Lepidoptera, II. Diptera, III. Coleoptera. *Memoirs of the Entomological Society of Canada* 78: 1-124.

Boucher, T.J. (2005). "Cucurbit Yellow Vine Disease (CYVD) In Connecticut". University of Connecticut, Cooperative Extension System. Retrieved 30 March 2016.

Britton, W.E. (1919) Insects attacking squash, cucumber, and allied plants in Connecticut. *Connecticut Agricultural Experiment Station Bulletin*. 216: 33-51.

Bruton, B.D., Mitchell, F., Fletcher, J., Pair, S.D., Wayadande, A., Melcher, U., Brady, J., Bextine, B. and Popham., T.W. (2003) *Serratia marcescens*, a phloem-colonizing, squash bug-transmitted bacterium: Causal agent of cucurbit yellow vine disease. *Plant Disease*. 87: 937–944.

Canhilal, R. and Carner. G.R. (2006). Efficacy of entomopathogenic nematodes (Rhabditida: Steinernematidae and Heterorhabditidae) against the squash vine borer, *Melittia cucurbitae* (Lepidoptera: Sesiidae) in South Carolina. *Journal of Agricultural and Urban Entomology*. 23: 27–39.

Canhilal, R. and Carner. G.R. (2007). *Bacillus thuringiensis*as a pest management tool for control of the squash vine borer, *Melittia cucurbitae* (Lepidoptera: Sesiidae) in South Carolina. *Journal of Plant Diseases and Protection*. 114: 26-29.

Canhilal, R.G., Carner, G.R. and. Griffin, R.P. (2006). Life history of the squash vine borer, *Melittia cucurbitae* (Harris) (Lepidoptera: Sesiidae) in South Carolina. *Journal of Agricultural and Urban Entomology* 23: 1–16.

Capinera, J. (1999). "Squah bug: Anasatristis". Featured Creatures. IFAS. Retrieved 29 March 2016.

Capinera, J.L. (2005). *Melon Worm. Diaphania hyalinata. Featured Creatures.* Division of Plant Industry, Department of Entomology and Nematology Florida Cooperative Extension Service, Institute of Food and Agricultural Sciences, University of Florida, Gainesville, FL. Publication # EENY-163.

Capinera, J.L. (2008). Spotted Cucumber Beetle or Southern Corn Rootworm, *Diabrotica undecimpunctata* Mannerheim (Coleoptera: *Chrysomelidae*). *Encyclopedia of Entomology* 3519–3522.

Davis, R.F. (2007). Effect of *Meloidogyne incognita* on watermelon yield. *Nematropica.* 37(2): 287–293.

Delahut, K. (2005). Squash vine borer. University of Wisconsin Garden Facts X1024. http://hort.uwex.edu/ articles/squash-vine-borer;

Dimetry, N.Z., El-Laithy, A.Y. AbdEl-Salam, A.M.E. and El-Saiedy, A.E. (2013). Management of the major piercing sucking pests infesting cucumber under plastic house conditions. *Archives of Phytopathology and Plant Protection.* 46(2): 158–171.

Diver, S. (2008). (Updated by Tammy Hinman). Cucumber beetles: Organic and biorational integrated pest management. ATTRA Publication #IP212. http:// attra.ncat.org/attra-pub/PDF/cucumberbeetle.pdf.

Ghosh, S.K. and Senapati, S.K. (2001). Biology and seasonal incidence of *Henosepilachna vigintioctopunctata* Fabr. on brinjal under terai region of West Bengal. *Indian Journal of Agricultural Research* 35(3): 149-154.

Godfrey, L.D. (2011). Pest notes: Spider mites. UC ANR Publication 7405. UC Statewide Integrated Pest Management Program, University of California, Davis, CA. http://www.ipm.ucdavis.edu/PMG/ PESTNOTES/pn7405.html.

Gupta, B.L. and Verma, A.K. (1995). Host specific demographic studies of the melon fruit fly, *Dacus cucurbitae* Coquillett (Diptera : *Tephritidae*). *Journal of Insect Science* 89(1): 87-89.

Hall, R.A. (1982). Control of whitefly, *Trialeurodes vaporariorum* and cotton aphid, Aphis gossypiiin glasshouses by two isolates of the fungus, *Verticillium lecanii*. *Annals of Applied Biology.* 101(1): 1–11.

https://pnwhandbooks.org/insect/vegetable/vegetable-pests/hosts-pests/cucumber-spider-mite.

https://www.google.com/search?q=red+pumpkin+beetle+on+cucumber& sxsrf=ACYBGNRE5HnNxEEGBrwvAI0GOjlTfHfwEQ:1577824880416

&source=lnms&tbm=isch&sa=X&ved=2ahUKEwjqgo_c3-DmAhUq6n
MBHR8yAb0Q_AUoAXoECBIQAw&biw=1920&bih=937 #imgdii=
7FG9Oz4fbboDzM:&imgrc=i5hHme-TlYu3GM:

https://www.google.com/search?q=red+pumpkin+beetle+on+cucumber
&sxsrf=ACYBGNRE5HnNxEEGBrwvAI0GOjlTfHfwEQ:
1577824880416&source=lnms&tbm=isch&sa=X&ved=2ahUKEwjqgo_c3-
DmAhUq6nMBHR8yAb0Q_AUoA XoECBIQAw&biw=1920&bih=937#
imgrc=nAyM3j70f1NQdM:

https://www.google.com/search?q=red+pumpkin+beetle+o
n+cucumber&sxsrf=ACYBGNRE5HnNxEEGBrwvAI0GOjlTfHfw
EQ:1577824880416&source=lnms&tbm=isch&sa=X&ved=
2ahUKEwjqgo_c3-DmAhUq6nMBHR8yAb0Q_AUoAXoECBIQ
Aw&biw=1920&bih=937#imgrc=T4shz8aQ3ed0MM:

https://www.google.com/search?q=red+pumpkin +beetle+on+cucumbe r&sxsrf
=ACYBGNRE5HnNxEEGBrwvAI0GOjlTfHfwEQ:1577824880416
&source=lnms&tbm=isch&sa=X&ved=2ahUKEwjqgo_c3-DmAhUq6n
MBHR8yAb0Q_AUoAXoECBIQAw&biw=1920&bih=937#imgrc=
nhDcggV4mBre8M:

https://www.google.com/search?biw=1920&bih=888&tbm=isch&sxsrf=AC
YBGNR4n5FBomK8WY7ynbSwgjuf71_8Qw%3A1577824883459&sa=1&ei=
c7ILXs_WG-aJ4-EPyfSIkA0&q=anadevidia+pe ponis&oq=anadevidia+
peponis&gs_l=img. 3..0l2.419802.42554 0..426674...0.0..0.176.3045.
0j19......0....1..gws-wiz-img.......35i39j0i67j0i131j0i10j0i30.rjgEoNyMN
f0&ved=0ahU KEwiP3Mjd3-DmAhXmxDgGHUk6AtIQ4dUDCAc&
uact=5#imgdii= Sl14p5uwl-p_oM:&imgrc=RAE_neWMzLVTpM:

https://www.google.com/search?biw=1920&bih =937&tbm=isch&sxsrf=
ACYBGNQ pe5iew68GNAO6M58q67N5v8oEeQ%3A15778 955865 19&s
a=1&ei=osYMXsO0H7HVz7sPnsK28AQ&q=bemisia+tabaci&oq=
bemisia+ta&gs_l=img.1.0.0l2j0i67j0l7.36658.39742..41581...0.0..0.
181.1738.0j11......0....1..gws-wiz-img.......35i39j0i131.A0gAjhTph-
g#imgrc=zDv_04WJzcB6oM:

https://www.google.com/search?biw=1920&bih=888&tbm=isch&sxsrf
=ACYBGNQoemj7uUCxE6b3l59UHSgKD6p2fw%3A1577895629077
&sa=1&ei=zcYMXoSkBODSz7sPvYefsAk&q=fruit+fly&oq=fruit+fly&gs_
l=img.3..0l10.114387.117030..117612...0.0..0.183.1595.0j10......0....1..gws-
wiz-img.......35i39j0i67.ODaZ452pvwo&ved=0ahUKEwjEvtqj5-
LmAhVg6XMBHb3DB5YQ4dUDCAc&uact=5#imgrc=Tde1

Islam, K., Islam, M.S. and Ferdousi, Z. (2011). Control of Epilachna
vigintioctopunctata Fab. (Coleoptera:Coccinellidae) using some indigenous plant
extracts. *Journal of Life and Earth Science.* 6: 75–80.

Kapoor, V.C. (2002). Fruit fly pests and their present status in India. In Proceeding of 6th International Fruit Fly Symposium,pp:23-33, Stellenbosch, South Africa.Retrieved fromhttps://nucleus.iaea.org/sites/naipc/twd/Documents/6thISFFEI_Proceedings/KAPOOR.pdf

Khan, M.M.H. (2012). Host preference of pumpkin beetle to cucurbits under field conditions. *Journal of the Asiatic Society of Bangladesh, Science.* 38(1):75-82.

Khan, M.M.H., Alam, M.Z. Rahman, M.M. Miah, M.I. and Hossain. M.M. (2012). Influence of weather factors on the incidence and distribution of red pumpkin beetle infesting cucurbits. *Bangladesh Journal of Agricultural Research.* 37(2): 361–367.

Khelu, A.Z., Zaets, V.G. and. Shesteperov. A.A. (1989). Tests on resistance to root-knot nematodes in tomato,cucumber and pepper varieties grown under cover in central Lebanon. *Byulleten' VsesoyuznogoInstituta Gel'mintologiiim. K.I. Skryabina.* 50: 85–89.

Klungness, L.M., Jang, E.B., Mau, R.F.L., Vargas, R.I., Sugano, J.S., Fujitani, E. (2005). New approaches to sanitation in a cropping system susceptible to tephritid fruit flies (Diptera: Tephritidae) in Hawaii. *Journal of Applied Science and Environmental Management.* 9: 5–15.

Koch, R.L. (2003). The multicolored Asian lady beetle, Harmoniaaxyridis: A review of its biology, uses in biological control, and non-target impacts. *Journal of Insect Science.* 3: 16–32.

Konstantopoulou, M. & Mazomenos, B.E. (2005). Evaluation of Beauveria bassiana and B. brongniartii strains and four wild-type fungal species against adults of Bactrocera oleae and Ceratitis capitata. *BioControl.* 5: 293-305.10.1007/s10526-004-0458-4.

Kumar, S. and Kumar, J. (1998). Laboratory evaluation of some insecticides against strains of hadda beetle (*Epilachna vigintioctopunctata* Fab.) resistant to malathion and endosulfan. *Pest Management and Economic Zoology,* 6(1-2): 133-137.

Laska, P., Perez-Banon, C., Mazanek, L., Rojo, S., Stahls, G., Marcos-Garcia, M.A., Bicik, V. and Dusek. J. (2006). Taxonomy of the genera Scaeva, SimosyrphusandIschiodon (Diptera: Syrphidae): Descriptions of immature stages and status of taxa. *European Journal of Entomology.* 103: 637–655.

Letourneau, D.K. (1986). Associational resistance in squash monocultures and polycultures in tropical Mexico. *Environmental Entomology.* 15: 285–292.

Lopez-Gomez, S. and Verdejo-Lucas. M. (2014). Penetration and reproduction of root-knot nematodes on cucurbit species. *European Journal of Plant Pathology.* 138: 863–871.

McAuslane, H.J. and Smith. H.A. (2000). Sweetpotato Whitefly B Biotype, Bemisiatabaci (Gennadius) (Insecta: Hemiptera: Aleyrodidae). EENY-129, University of Florida/IFAS Extension. University of Florida, Gainesville, FL. https://edis.ifas.ufl.edu/in286.

McKinlay, R.G. (1992). Vegetable Crop Pests. Boston, MA: CRC Press : 98–101.

McSorley, R. and Waddill, V.H. (1982). Partitioning yield loss on yellow squash into nematode and insect components. *Journal of Nematology*. 14: 110–118.

Messenger, B. and Braun. A. (2000). Alternatives to Methyl Bromide for the Control of SoilBorne Diseases and Pests in California. Pest Management Analysis and Planning Program. http://cdpr.ca.gov/docs/emon/methbrom/alt-anal/sept2000.pdf

Mitchell, R.F. and Hanks. L.M. (2009). Insect frass as a pathway for transmission of bacterial wilt of cucurbits. *Environmental Entomology.* 38(2): 395-403.

Mohaned, M.A., Mohamed, M., Elabdeen, H.Z. and Ali, S.A. (2013). Host Preference of the Melon Worm, DiaphaniahyalinataL. (Lepidoptera: Pyralidae), on Cucurbits in Gezira State, Sudan. *Persian Gulf Crop Protection*. 2(3): 55-63.

Mohasin, M. and De, B.K. (1994). Control *Henosepilachna vigintioctopunctata* Fabr. On potato in West Bengal plains. *Journal of the Indian Potato Association.* 21(1-2): 151-153.

Mondal, S., and. Ghatak. S.S. (2009). Bioefficacy of some indigenous plant extracts against epilachna beetle (HenosepilachnavigintioctopunctataFabr.) infesting cucumber. *Journal of Plant Protection Sciences.* 1(1): 71-75.

Mukhtar T., Kayani M.Z., Hussain, M.A. (2013). Response of selected cucumber cultivars to *Meloidogyne incognita. Crop Protction.* 44: 13-17.

Murphy, G., Ferguson, G. and Shipp. L. (2014). Mite pests in greenhouse crops: Description, biology and management. www.omafra.gov.on.ca/english/crops/facts/14-013.html.

Noling, J.W. (2009). Nematode management in cucurbits (cucumbers, melons, squash). http://edis.ifas.ufl.edu/ ng025.

Nyoike, T.W. (2007). Evaluation of living and synthetic mulches with and without a reduced-risk insecticide for suppression of whiteflies and aphids, and insect transmitted viral diseases in zucchini squash. A thesis submitted to the Graduate school in partial fulfillment for MS in integrated pest management. University of Florida, Gainesville, FL, p. 90.

Park, Y.L. and Lee, J.H. (2005). Impact of two spotted spider mite (Acari: Tetranychidae) on growth and productivity of glasshouse cucumbers. *Journal of Economic Entomology*. 98: 457–463.

Rahaman, M.A. and Prodhan. M.D.H. (2007). Effects of net barrier and synthetic pesticides on red pumpkin beetle and yield of cucumber. *International Journal of Sustainable Crop Production*. 2(3): 30-34.

Rajagopal, D. and Trivedi, P. (1989). Status, bioecology and management of epliachna beetle, *Epilachna vigintioctopunctata* (Fab.) (Coleoptera: Coccinellidae) on potato in India. *Tropical Pest Management*. 35(4): 410-413.

Rath, L.K., Nayak, V.S. and Dash, D. (2002). Non-preference mechanism of resistance in egg plant to epliachna beetle, *Henosepilachna vigintioctopunctata*. *Indian Journal of Entomology*. 64 (1): 44-47.

Sapkota, R., Dahal, K.C. and Thapa, R.B. (2010). Damage assessment and management of cucurbit fruit fliesin spring-summer squash Use of pheromone traps. *Journal of Entomology and Nematology*. 2(1): 7-12.

Shooker, P., Khayrattee, F. and Permalloo, S. (2006). Use of maize as a trap crops for the control of melon fly, B. cucurbitae (Diptera:Tephritidae) with GF-120. Bio-control and other control methods [Online]. Available on: http\\www.fcla. edu/FlaEnt/fe87 p354.pdf. [Retrieved on: 20th Jan. 2008].

Smith, R.I. (1911). Two important cantaloupe pests. *North Carolina Agricultural Experiment Station Bulletin*. 214: 101-146.

Snyder, W.E. (2015). Managing cucumber beetles in organic farming systems. Department of Entomology, Washington State University Pullman, WA. (Accessed 04/10/2018).

Sorensen, K.A. (1999). Cucumber beetles, Coleoptera: Chryso¬melidae. Greenshare fact sheets. University of Rhode Island Landscape Horticulture Program. (12 Sep 2013).

Stapleton, J.J. and Summers. C.G. (2002). Reflective mulches for management of aphids and aphid-borne virus diseases in late-season cantaloupe (Cucumis melo L. var. cantalupensis). *Crop Protection*. 21: 891–898.

Synder, W. (2012). Managing cucumber beetles in organic farming systems. Cornell University Corporative Extension. (12 Sep 2013).

Venkataraman, T.V., Mathur, V.K. and Ramesh, C. (1962). Experiments on the possible use of Bacillus thuringiensis in the control of crop pests. *Indian Journal of Entomology* 24 : 274-277.

Wang, K.H. and McSorely. R. (2004). Management of Nematodes with Cowpea Cover Crops. UF IFAS Extension, ENY-712. Electronic. http://edis.ifas.ufl.edu/pdffiles/IN/IN51600.pdf.

Webb, S. (2010). Insect management for cucurbits (Cucum¬ber, Squash, Cantaloupe, and Watermelon). ENY-460. Gainesville: University of Florida Institute of Food and Agricultural Sciences (12 Sep 2013).

Welty, C. (2009). Squash vine borer. Agriculture and Natural Resources Fact Sheet HYG-2153-09. The Ohio State University Extension.

Westerdahl, B.B. and Becker, J.O. (2011). UC IPM Pest Management Guidelines: Cucurbits. http://www.ipm.ucdavis.edu/PMG/r116200111.html.

Worthley, H.N. (1923). The squash bug in Massachusetts. *Journal of Economic Entomology.* 16: 73–79.

Xalxo, P.C., Karkun, D. and Poddar. A.N. (2013). Rhizospheric fungal associations of root knot nematode infested cucurbits: In vitro assessment of their nematicidal potential. *Research Journal of Microbiology.* 8: 81–91.

Yardim, E.N. Arancon, N.Q., Edwards, C.A., Oliver, T.J. and Byrne, R.J. (2006). Suppression of tomato hornworm (Manducaquinquemaculata) and cucumber beetles (AcalymmavittatumandDiabotricaundecimpunctata) populations and damage by vermicomposts. *Pedobiologia.* 50: 23-29.

Zehnder, G. (2011). Biology and management of pickleworm and melonworm in organic curcurbit production systems. Electronic. http://www.extension.org/ pages /60954/biology-and- management-of-picklewormand- melonworm-in- organic-curcurbit- production- systems#. VNmbv 3uH6yc.

13

Biointensive Integrated Pest Management of Tomato

Dhole, R.R, Patil, S.P. Maurya, R. and Singh, R.N.

Introduction

Tomato (*Solanum lycopersicum*) is the largest and most important vegetable crop belongs to family Solanaceae grown all over the world. It is grown for its edible, fleshy, berry typed, mostly red coloured fruits.The species has its origin in coast of Ecuador, Peru, part of Bolivia and the northern region of Chile i.e.western South America and cultivation started most probably by the people of Maxico of Central America[1-3]. It is found to be a rich source of many minerals like Ca, K, Cu, Fe, Zn and S and vitamins and other nutritive values so it is cultivated both as protective and supplementary food[4].In India and in some European countries with common form of tomato fruits the cherry type of tomato fruits also cultivated on large scale. Cherry type of tomato fruits when consumed fresh, act as a very efficient source of diet of urban and rural people as it is reported to be a good source of many macro elements of human requirements such as K, Mg, P and Na[5]. As it is short duration crop and gives high yield, it is important from economic point of view and hence area under its cultivation is increasing day by day. India is considered to be a second largest producer of tomatoes succeeding to China. Tomato is used in preserved products like ketchup, sauce, chutney, soup, paste, puree etc. Cherry fruits are commonly used in diverse ways, including raw in salads and processed, and their acidity allows for ease of preservation without refrigeration (long shelf life) or home canning processing for later use[6].

India possesses all the required climatic conditions for the good growth and development of tomato and its produce. Instead of this farmer is not getting the appropriate return in terms of production and yield. There are many obstacles and biotic factors which hinder the rate of production of tomato crop. Among them most important are the insect pests and their incidence and impact on tomato crop which is responsible for major yield loss of tomato produce. Due to attack of insect pest, crop plant and its produce suffer a lot, many sucking pest acts as a vector to transmit the virus and infect the plant with viral diseases, it also causes crinkling, deforming of plant leaves which indirectly reduce down the rate of photosynthesis which further leads to decrease in total berry biomass production. Similarly, the attack by many borer type insects causes large scale damage to crop plant by making hole in stems, leaves, fruiting bodies etc. and it also reduce down the shelf life of produce.

Some major insect pest infesting tomato crop with their characteristics features like host range, distribution, seasonal incidence, biology and life cycle, symptoms of damage and their Biointensive integrated pest management techniques are being discuss in this chapter.

Major insect pest of tomato

The major insect pests of tomato crop are discussed along with their Biointensive Integrated Pest Management strategies and they can be classified based on their feeding habits as fruit and foliage feeders.

Insect pests	Scientific name	Family	Order
A. Fruit Borer			
Pod Borer	*Helicoverpa armigera* (Hübner)	Noctuidae	Lepidoptera
B. Foliage Feeder/ Leaf Minor			
Tobacco caterpillar	*Spodoptera litura* (Fabricius)	Noctuidae	Lepidoptera
Serpentine leaf miner	*Liriomyza trifolii* (Burgess)	Agromyzidae	Diptera
C. Sucking Insect Pests			
Whitefly	*Bemisia tabaci* (Gennadius)	Aleyrodidae	Hemiptera
Thrips	*Thrips tabaci, Frankliniella schultzei*	Thripidae	Thysanoptera
Spider mite	*Tetranychus* spp.	Tetranychidae	Acarina

1. Pod borer / Tomato Fruit Borer, *Helicoverpa armigera* (Hübner)

Systemic Position

Kingdom	:	Animalia
Phylum	:	Arthropoda
Class	:	Insecta
Order	:	Lepidoptera
Super family	:	Noctuoidea
Family	:	Noctuidae
Genus	:	*Helicoverpa*
Species	:	*armigera*

Host range

Pod borer, *Helicoverpa armigera* (Hübner) is a polyphagous pest of many crops in many parts of the world and is reported to infest more than 60 plant species of 47 families[10-11]which primarily includes cotton (*Gossypium hirsutum* L.), corn (*Zea mays* L.), tomato (*Lycopersicon esculentum* Mill), hot pepper (*Capsicun frutescens* L.), tobacco (*Nicotiana tobacum* L.), common bean (*Phaseolus vulgaris* L.), sunflower (*Helianthus annuus* L.) pigeon pea (*Cajanus cajan*), chickpea (*Cicer arietinum*), rice (*Oryza sativa*), sorghum (*Sorghum bicolor*) cowpea (*Vigna unguiculata*) etc.[8]

Distribution

This lepidopteran pest is distributed all over the world i.e. eastwards from Southern Europe and Africa through the Indian subcontinent to Southeast Asia, and spread up to China, Japan, Australia and the Pacific Islands[12]. It is mostly occurred from all parts of India like Madhya Pradesh, Assam, Orissa, West Bengal, Delhi, Haryana, Himachal Pradesh, Rajasthan, Punjab, Uttar Pradesh, Gujarat, Maharashtra, Andhra Pradesh, Karnataka, Tamil Nadu etc.

Symptoms of damage

Helicoverpa is considered as a pest of many crop plants because of its four characteristics (polyphagy, high mobility, high fecundity and a facultative diapause) which it shows during different stages of life cycle that enable it to survive, infest and feed in unfavourable habitats and adapt to abrupt climatic conditions[19]. Direct damage of the larvae of this noctuid pest to flowering and fruiting structures together with spitted out excreta which favours the fungal pathogen attack results in low crop yield and high costs of production. Larval stage is considered as most damaging stage in most of the crops including tomato.

Seasonal incidence

In *Kharif* season, under protected cultivation the incidence of pod borer *Helicoverpa armigera* (Hübner) starts from 35th meteorological week (last week of August to 1st week of September) and remain up to 49th meteorological week (last week of November to 1st week of December) and maximum population in field occurred in the month of November[20]. In *Rabi* Season, the incidence of it starts from the December month only and remain up to the harvest of crop (March-April) and peak period of incidence found in the last fortnight of March[21].

Biology

Pod borer *Helicoverpa armigera* (Hübner) female can lay several hundred white to greenish spherical eggs of 0.4 to 0.6 mm in diameter, on different plant parts. Hatching of eggs could takes place within 3-5 days after adults had laid eggs. As larva emerges out of egg, it bears a yellow head with numerous dots on it. The dorsal-side colour of larvae is mostly greenish and yellow to red-brown & it needs 13 to 22 days to develop fully[22].

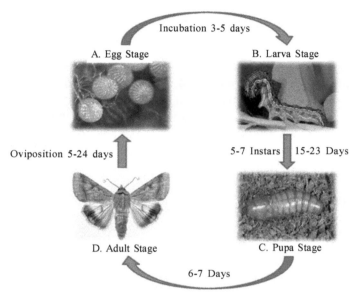

Fig. Diagrammatic representaton of life cycle of Tomato Fruit Borer, *Helicoverpa armigera* Hubn.

The characteristics feature of larva of *H. armigera* is the presence of three dark stripes which extended along the dorsal side and one yellow light stripe which is situated under the spiracles on the lateral side. The ventral parts of the larvae are pale. The development of pupae takes place inside the silken cocoon for 10 to 15 days in soil at a depth of 4–10 cm (1.6–3.9 inch). The developmental time of immature stages on tomato is approx. 35.1 days. Many researchers had concluded that *H. armigera* can complete their life cycle on all host plants, although tomato and hot pepper were relatively unsuitable[8]. On an average, total life cycle completed in 40-59 days[23].

Biointensive Integrated/Insect Pest Management

To control the infestation of *Helicoverpa*, chemical control measures are mostly preferred. This includesthe use of synthetic insecticides including organophosphates, synthetic pyrethroids and biorational com pounds are the main method for control of *H. armigera* in different parts of the world. This wide use of pesticides is of environmental concern and has repeatedly led to the development of pesticide resistance in this pest. Furthermore, the deleterious effects of insecticides on non-target organisms including natural enemies are among the major causes of pest out breaks[18]. It is therefore necessary to develop a novel strategy to manage population of *H. armigera* and reduce the hazard of of synthetic chemicals. For sustainable agricultural growth and environmental safety, use of biopesticides can play a vital role[17]. Biopesticides to be used, include:

A. Plant origin pesticides/ Botanical pesticides

B. Microbial (Viral, bacterial, protozoan and fungal),

C. Entomopathogenic nematodes (EPNs)

D. Predators and parasitoids

Garlic chilli kerosene extract (GCKE) @ 25 l/ha, Azadirachtin 3% WSP (400 g/ ha), Neem Seed Kernal Extract (NSKE) @ 25 l/ha etc. are reported to be most effective botanical pesticides for the control of tomato borer, *H. armigera*[15].

Bacillus thuringenesis (Bt) is known and most reported microbial type of biological control. It infest against gut of larva of *H. armigera*and reduce down its feeding potential and simultaneously results into death of insect pest[18].

HaNPV @ 10-10³polyhedra/ml is a Potent virus-based pesticides for management of *H. armigera*[16].

Steinernema carpocaspe @ 5x10⁹ ijs/ha, *S. riobrae*@ 5x10⁹ ijs/ha are the major potential EPN found effective against *H. armigera*. Similarly, application of *Heterorhabditis megidis and H. Indica* @ 800 ijs/ml and 5x10⁹ ijs/ha also known to reduce down the population of *H. armigera* in the field[18].

Weekly release of egg parasitoids *Trichogramma chilonis* @ 150,000 per hector (approx. 6 times) starting from 40th days after planting or during egg-laying period is reported to be most beneficial against *H. armigera*. Similarly release of *Trichogramma brasiliense, T. pretiosum* and *T. chilonis* @ 50,000/ha. in tomato crop at weekly interval from 25th days after planting gives most promising results for control of *H. armigera*[15-18].

2. Tobacco caterpillar: *Spodoptera litura* (Fabricius)

Systemic Position

Kingdom	:	Animalia
Phylum	:	Arthropoda

Class	:	Insecta
Order	:	Lepidoptera
Super family	:	Noctuoidea
Family	:	Noctuidae
Genus	:	*Spodoptera*
Species	:	*litura*

Host range

Spodoptera litura is a notorious leaf feeding insect pest of more than one hundred plants around the Asia-Pacific region. Host plant survey for this insect pest revealed 27 plant species as host plants of *S. litura* belonging to 25 genera of 14 families including cultivated crops, vegetables, weeds, fruits and ornamental plants[24]. Major host plants on which it thrived for maximum period were Cotton (*Gossypium hirsutum* L.), Castor (*Ricinus communis* L.), Cabbage (*Brassica oleracea* var. *botrytis* L.), Taro (*Colocasia esculenta* L.), Pigweed (*Trianthema portulacastrum* L.) and Sesban (*Sesbania sesban* L.)[24], tobacco (*Nicotiana tabacum*), Chinese cabbage (*Brassica rapa* subsp. pekinensis), cowpea (*Vigna unguiculata*) and sweet potato (*Ipomoea batatas*)[25] etc. There are 112 reported host plant of *S. litura*[29-30] out of which 60 host plants had been reported in India[31].

Distribution

The common cutworm, *Spodoptera litura*, is a pest of many crops across globe throughout tropical and temperate Asia, Australia and the Pacific Islands and has become a major pest of soybean (*Glycine max*) throughout its Indian range[27-28]. It occurs mostly in the area having temperature range of 15-36°C[26]. In India, the incidence of this pest occurs in nearly all states but the predominant infestation is reported in Orissa, Karnataka, Maharashtra, Rajasthan, Andhra Pradesh, Tamil Nadu, Uttar Pradesh, Madhya Pradesh etc.[32-34]

Symptoms of damage

A single larva per square metre is reported to cause average pod yield loss of 27.3% in groundnut through damage to various plant parts like leaves, flowers and pods[38]. The larva has vigorous eating patterns mostly included scraping of leaves, many times leaving the leaves only with its mid rib which directly decrease down the area of photosynthesis and hence reduce down the yielding potential of crop. The moth's effects are quite disastrous, destroying economically important agricultural crops and decreasing yield in some plants completely[38].

Seasonal incidence

In *Kharif* season, the pest remains active from end of July or mid of August to October coinciding with warm and humid climate and peak reproductive phases of soybean, causing 26–29% yield losses[33]. In Rabi/Summer Season the pest incidence starts from 5th meteorological week onwards means from the last week of January and its peak period is reported in 11-12 meteorological week i.e. last fortnight of March and remain up to harvesting period[41].

Biology

It is true that, life cycle of *S. litura* varies from host to host and other environmental and topographic conditions but an average a typical *S. litura* completes nearly 11-12 generations per year. Being poikilothermic organisms, the developmental rate in insects is highly contingent on external temperature conditions. Hence, temperature is generally considered the single most significant environmental factor influencing behaviour, distribution, development, survival and reproduction in insects[40]. Females could lay no eggs at the extreme low (15°C) and high (>35°C) test temperatures, demonstrating the importance of optimum temperature in determining the suitability of climate for the mating and reproduction in *S. litura*[26].

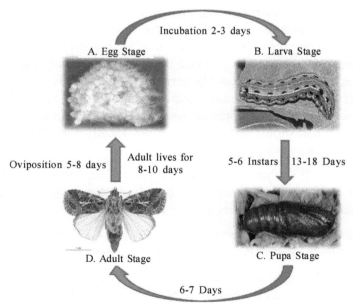

Fig. Diagrammatic representaton of life cycle of Tobacco caterpillar, *Spodoptera litura* (Fabricius)

At optimum conditions, adult female lays spherical and slightly flattened orange-brown or pink colour egg on the surface of leaves in big batches/masses (4-7 mm), with each cluster usually containing several hundred eggs. Average fecundity power of female is 2000 to 2600 eggs. When laid, the egg batches resembles as a golden brown balls as they are covered with hair scales provided by the female and they hatch within 2-3 days. The younger lighter green coloured larva of 2.3 mm size comes out as egg hatched and with the time convert into older dark green or brown coloured larva of 32 mm. Presence of Yellow Strip along the mid dorsal portion is a characteristic feature of the larvae. Being a nocturnal pest, it feeds generally in nights. During whole larval period *S. litura* passes through six instars. Pupation takes place in the soil near the base of the plant and lasts for 7 to 10 days. Pupa is of red brown coloured with two small spines at the tip of the abdomen[37,39].

A grey-brown coloured moth is emerged out from the pupa which is approx. 15-20 mm long and with a wingspan of 30-38 mm. Dark grey, red and brown coloured pattern can be observed on forewings while hind wings are greyish-white with a grey outline. Adult female survive for 8-9 days while survival of male adults varies from 10-11 days. [37,39]

Biointensive Integrated Pest Management

In case of *S. litura* many Biointensive pest management strategies had been widely used all over the world which is known to give the appropriate results with no health hazards to environment, animals and human beings. Biointensive pest management prevent and reduce down the unnecessary use of chemical pesticides hence the input cost in agriculture and so this pest management strategies are widely accepted, economically feasible and politically promoted one. It includes the use of botanical pesticides, microbial products and natural enemies i.e. parasitoids and predator. Some mostly used Biointensive pest management strategies are as follow:

SlNPV @ 10-10³ polyhedra/ml and *Bacillus thuringiensis* serotype *kurstaki* (Btk) @ 0.032% is a potent virus-based and bacterial based pesticides recommended for management of *S. litura*[43].

Steinernema carpocaspe @ 5x10⁹ ijs/ha, *S. glaseri* @ 5x10⁹ ijs/ha are the major potential EPN found effective against *S. litura*.[18]

The most widely used egg parasitoid against *S. litura* is *Telenomus remus*@ 120000/ha and it is recommended to use 5 times at weekly intervals[18]

Azadirachtin 3% WSP (400 g/ha), Neem Seed Kernal Extract (NSKE) @ 25 l/ha, Neem Oil 5% etc. are reported to be most effective botanical pesticides[15,42].

3. American serpentine leaf miner, *Liriomyza trifolii*

Systemic Position

Kingdom	:	Animalia
Phylum	:	Arthropoda
Class	:	Insecta
Order	:	Diptera
Section	:	Schizophora
Family	:	Agromyzidae
Genus	:	*Liriomyza*
Species	:	*trifolii/brasslike*

Host range

This is primarily the pest of chrysanthemums (*Chrysanthemum indicum*) and celery (*Apium graveolens*) but it has a wide host rangealso including bean (*Phaseolus vulgaris*), beet (*Beta vulgaris*), carrot (*Daucus carota*), cucumber (*Cucumis sativus*), eggplant (*Solanum melongena*), lettuce (*Lactuca sativa*), melon (*Cucumis melo*), onion (*Allium cepa*), pea (*Pisum sativum*), pepper (*Piper nigrum*), potato (*Solanum tuberosum*), cotton (*Gossypium hirsutum* L.), tomato (*Lycopersicon esculentum* Mill.) and cowpea (*Vigna sinensis* (L.) Walp) etc.[48-49] It is also reported to be a pest of many flower crops like chrysanthemum, gerbera, gypsophila and marigold, and also the crops from compositae family[56-58].

Distribution

Theleafminer flies are reported as a quarantine pest in many countries. The genus *Liriomyza* exhibits cosmopolitan distribution and hence observed worldwideprincipally in tropical and subtropical regions for an instance from eastern United States and Canada, northern South America to Europe and eastern southern Asia and found to be associated with several host plants[58].

Symptoms of damage

The female of *Liriomyza* makes numerous punctures to the leaf mesophyll with her ovipositor mostly during daylight hours which gives stippled appearance on foliage, and uses these punctures for feeding and egg laying[54]. The larva of *Liriomyza* irregularly mine out the leaves of crop with blotch-like endings, which results in destruction mesophyll of leaves and indirectly reduce down the rate of photosynthesis[52]. Extensive mining also causes premature leaf drop, which can result in lack of shading and sun scalding of fruit. Wounding of the foliage also allows entry of bacterial and fungal diseases[55]

Seasonal incidence

The *Liriomyza* starts to observe and give its first appearance in the field of *Rabi* tomato in 6th-8th and 9th standard meteorological week (SMW) i.e. (February and March), attain its peak population in 14th and 17th SMW (April)[59-60]. In Kharif season, the incidence of this pest occurred in 38-40th standard meteorological week (SMW) with its initial mark in the month of August and its peak population in month of October[61].

Biology

The female of *Liriomyza* lays the oval shaped, small sized eggs by inserting them in little below the surface of epidermis on the underside of the leaves which hatches out in 3-5 days[45]. The oviposition occurred at a rate of 35 to 39 eggs per day, for a total fecundity of 200 to 400 eggs.

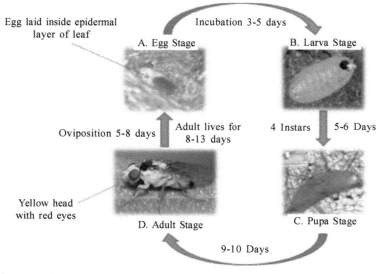

Fig. Diagrammatic representaton of life cycle of American serpentine leaf miner, *Liriomyza trifolli*

The larva of *Liriomyza* is legless with three active larval instars and dynamic variation in body colour i.e. from colourless to yellowish colour. The different parameters of the body and mouthparts can be used to differentiate instars; the latter is particularly useful[48]. A fourth instar occurs between puparium formation and pupation, where larva drops to the soil to pupate. The colour and size of puparium varies from initial yellowish to darker brown and oval and slightly flattened ventrally, respectively. The reported average age of puparium is 9.3 days[51]. Adults are small, measuring less than 2 mm in length, with a transparent wing span of 1.25 to 1.9 mm. The head is yellow with red eyes. The thorax and abdomen are mostly grey and black although the ventral surface and legs are yellow. Adults live about 8-13 days. The life cycle of *Liriomyza* is short i.e. of 21 to 28 days[58].

Biointensive Integrated/Insect Pest Management

Tobacco extract (5%), Azadirachtin 3% WSP (400 g/ha), Neem Seed Kernal Extract (NSKE) @ 25 l/ha etc. are reported to be most effective botanical pesticides for the control of American serpentine leaf miner, *Liriomyza trifolii*[46,47,50].

Bacillus thuringenesis var *Kurstaki* (Btk) is known and most reported microbial type of biological control against *Liriomyza trifolii*[53].

Ha NPV @ 250 LE/ha DFP is a Potent virus-based pesticides for management of *Liriomyza trifolii*[54]. *Photorhabdus luminescens* a gamma proteobacterium which lives in the gut of an entomopathogenic nematode of the family Heterorhabditidae is also reported to be lethal against *Liriomyza trifolii*.

Steinernema carpocaspe, S. riobrae, S. feltiae (Filipjev) @ 5x10⁹ ijs/ha are the major potential EPN found effective against *H. armigera*. Similarly, application of *Heterorhabditis megidis and H. Indica* @ 800 ijs/ml and 5x10⁹ ijs/ha also known to reduce down the population of *Liriomyza trifolii* in the field[44,64].

The parasitoids *Diglyphus isaea, D. begini, D. sibirica* @ 0.13 females/plant for 5 releases, 0.19 females/plant for 8 releases and 0.15 females/plant for 3 releases from spring through summeris reported to be most beneficial against *H. armigera*. Similarly 5 releases of *Hemiptarsenus varicornis* on cherry tomatoes in greenhouses at rates of 0.33 and 0.16 females per plant over 2 months gives good control of *Liriomyza trifolii*. Other notable parasitoids for control of *L. trifolii* are *Opius pallipes, Neochrysocharis formosa* etc[63,65].

4. Whitefly: *Bemisia tabaci* (Gennadius)

Systemic Position

Kingdom	:	Animalia
Phylum	:	Arthropoda
Class	:	Insecta
Order	:	Hemiptera
Superfamily	:	Aleyrodoidea
Family	:	Aleyrodidae
Genus	:	*Bemisia*
Species	:	*tabaci*

Host range

It feed on several field crops in tropical and sub-tropical countries like tomato, tobacco, cassava, cabbage, cauliflower, melon, mustard, brinjal, cotton etc. Many indigenous obscure weed species are the host of this pest[84].

Distribution

Bemisia tabaci is a cosmopolitan found on over 900 host plants on all continents except Antarctica[73].In India, it is found nearly in most of the states with more or lesser damaging potential[73].

Symptoms of Damage

Damaging stage of this pest is both nymph and adults which causes damage mainly in three ways viz; it act as vector for transmitting a number of viral diseases including leaf curl virus, it lower down the plant strength and vitality by reduce down the plant sap, it interfere with normal photosynthesisbecause of growth of sooty mould on the honey dew excreted by the it. It not only suck the plant sap while feeding but also transmit tomato leaf curl virus (TLCV), which results in curling of tomato leaves[70].

Seasonal incidence

Whitely start to appear in field of tomato during *Rabi* season from 50th SMW i.e. second week of December itself and its peak population incidence is generally observed in the second week of March i.e. 11th SMW[66]. Sometimes in some areas maximum incidence of whitefly starts from the January itself and varies from place to place[67]. In Summer season crop (March to June), the incidence of whitefly starts to appear just after a fortnight interval of transplanting the seedling to main field that is in 13th SMW and reached to its peak level in 21st SMW[68]. Adult whitefly present throughout the growing period in the tomato field and their population was highest at the end of rainy season. It starts to appear from the 32nd SMW and reached at peak at 37-38th SMW[69,71,72].

Biology

Adult female can lay more than 200 whitish to brown eggs in circular manner, on lower surface of leaves having long axis perpendicular to leaf which hatch out on 5-10 days[82].

The first instar nymph is flat, oval and scale-like having length of 0.3 mm and is the only mobile stage (called the "crawler"). It moves on leaf surface in search of a feeding location where it can get in good touch with phloem then it moults, become sessile until all the remaining nymphal stages pass out. Initial three nymphal stages are of 3-4 days duration and 4th nymphal stage is known as puparium which last for 6 days. All stages feed on intracellular liquids[83].

With T-shaped rupture the adults emerge out from the puparium, and spread out its wings completely and dust them with wax from the gland of abdomen. The intercourse begins after 14-20 hours of emergence and can happen for several times. Female can live for longer duration of time i.e. 60-65 days than male i.e. 9-17 days. On an average, 11-15 life cycles can be completed in a year[84-85].

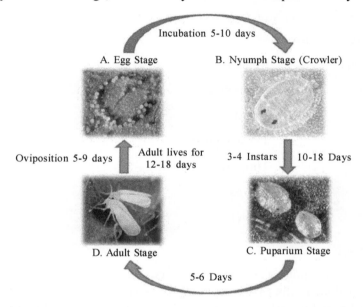

Fig. Diagrammatic representaton of life cycle of White fly, *Bemisia tabaci* (Gennadius)

Biointensive Integrated/Insect Pest Management

Azadirachtin 3% WSP (400 g/ha), Neem Seed Kernal Extract (NSKE) @ 25 l/ha, Tobacco extract (5%), Garlic extract etc. are some botanical formulations which are found to be effective for control of incidence of whiteflies[86].

There are many predators reported as a natural enemies of whitefly like *Amblyseius limonicus, A. swirskii, Transeius montdorensis* which are responsible for reduce down and monitor the pest population of it in field conditions[74].

Many parasitoids also act as effective natural enemies of whiteflies like *Encarsia Formosa, Eretmocerus corni, Eretmocerus mundus, Isaria fumosorosea* etc.[75,76]

Steinernema feltiae, S. carpocapsea etc. are the reported entomopathogenic nematodes for the control and monitoring the population of whitefly[77-78].

Lecanicillium muscarium, *Isaria fumosorosea*and*Beauveria bassiana* are the entomopathogenic fungi which are most widely used in eradication programme of *Bemesia* tabaci from crop ecosystem[79-80].

Amblyseius swirskii, *Transeius montdorensis* and *Typhlodromalus limonicus* are the reported predatory mites having wide range of efficacy against various lifestages of *B. tabaci*[81].

5. Thrips: *Thrips tabaci*

Systemic Position

Kingdom	:	Animalia
Phylum	:	Arthropoda
Class	:	Insecta
Order	:	Thysanoptera
Family	:	Thripidae
Genus	:	*Thrips*
Species	:	*tabaci*

Host range

Thrips is a cosmopolitan pest and found to be infesting on many host crops in the world including alfalfa (*Medicago sativa*), asparagus (*Asparagus officinalis*), bean (*Phaseolus vulgaris*), beet (*Beta vulgaris*), blackberry (*Rubus fruticosus*), cabbage (*Brassica oleracea*), carrot (*Daucus carota*), cauliflower (*Brassica oleracea* L. var. *botrytis* L.), celery (*Apium graveolens*), cotton (*Gossypium* spp.), cucumber (*Cucumis sativus*), garlic (*Allium sativum*), kale (*Brassica alboglabra* L.H. Bailey), leek (*Alliumampeloprasum* var. porrum), lettuce (*Lactuca sativa*), onion, parsley (*Petroselinum crispum*), pea (*Pisum sativum*), pineapple (*Ananas comosus*), potato (*Solanum tuberosum*), pumpkin (*Cucurbita maxima*), squash (*Cucurbita* spp.), strawberry (*Fragaria ananassa*), sweet potato (*Ipomoea batatas*), turnip (*Brassica rapa* var. rapa), tomato (*Solanum lycopersicum*) etc. and practically all small grains[87-88]. Many grass and weeds are also reported to be a well-known host of it[89-90].

Distribution

As already described above thrips is a cosmopolitan pest and found mostly in every part of world. It is predominantly noticed in Ontario, Australia, Canada, New York, New Zealand, Europe, Asia and Some parts of Africa[98].

Symptoms of Damage

The nymph and adult of thrips and lacerate and punctures the surface tissues of the foliage and suck the sap fluid from it[88]. The infestations by thrips results in the appearance of spots and white blotches on leaves and periodically leaves dries up and get fall down[92].

Seasonal incidence

In *kharif* season, the appearance of thrips in field starts from the 25th SMW i.e. 1st fortnight of July and attain its peak population from 32nd to 35th SMW i.e. 1st fortnight of September[96]. The population of thrips in field found to be highest in the month of Sep-Oct and starts to decline in the month of November[95].

Similarly, in *Rabi* season, the incidence starts with 46th SMW and peak period of thrips population generally observed in the month of February and March[97].

Biology

The female of it can reproduce 50-60 to 150-300 white-bean shaped eggs with reference to host plant and environmental conditions both with mating and without mating and the eggs laid singly and parthenogenetically in absence of male[93].

On average 4-9 days are needed for hatching of eggs and nymph comes out of eggs which are very much similar to the adults. They start feeding immediately after their emergence by lacerating the epidermal tissues of the leaves and swallowing the sap of the host[94].

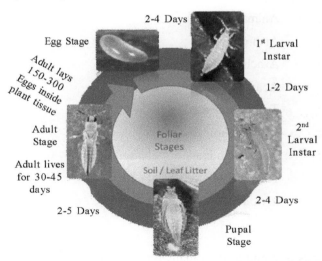

Fig. Diagrammatic representaton of life cycle of Thrips, *Thrips tabaci*

This pest bears two pupal stages: Prepupal and pupal stages which it spends at a depth of about 2.5 to 5.0 cm inside the soil[93].

The body of adult is small (1mm) and yellowish-brown in colour, with slender thorax and posteriorly tapering abdomen. Female bears full wings but in male the wings are extremely reduced or absent. Adult female lives for 2-4 weeks[89,92,93].

Biointensive Integrated/Insect Pest Management

The nymph and adults are damaging stages of thrips and are difficult to control using classical biological control due to their small sizes and high rates of reproduction[93]. Only two families of parasitoid the Eulophidae and the Trichogrammatidae of Hymenoptera are reported to be effective against it. Anthocorid bugs of genus *Orius* and phytoseiid mites are the other known predators. EPF like *Beauveria bassiana* and *Verticillium lecanii* can kill thrips at all life-cycle stages[94].

Many botanicals are also reported to give better result in controlling of thrips population in field like neem seed powder extract, neem soap and essential oils from basil or tulsi[99]. Similarly, the extracts from different parts of many pants *Azadirachta indica, Chrysanthemum cinerariaefolium, Cymbopogon citrates, Nicotiana tabacum, Parthenium hysterophorous* serves as effective pesticide to monitor the incidence of thrips in field conditions[100].

6. Spider mite: *Tetranychus* spp.

Systemic Position

Kingdom	:	Animalia
Phylum	:	Arthropoda
Class	:	Arachnida
Subclass	:	Acari
Order	:	Trombidiformes
Superfamily	:	Tetranychoidea
Family	:	Tetranychidae
Genus	:	*Tetranychus*
Species	:	*urticae*

Host range

It is a polyphagous pest and feed on many numbers of plants, weeds and forest trees[108,110].

Distribution

This pest is having vast geographical spred over across all Mediterranean region like North America, Eurasia, Africa Europe and Japan which has the potential to be extensively colonised by the mite population[111].

Symptoms of Damage

The pest causes damage by perforating lower epidermal cells. High infestation of these mites reduce the photosynthetic rate, damage the leaf mesophyll and cause closure of stomata. This leads to the formation a chlorotic spot[105]. The spider mites generally feed on the lower surface of the leaves as a result the infested leaves initially show speckling and later turn yellowish, finally leading to defoliation. The mites spread to all parts of the plants as the population increases especially during day periods and produce webbing over the entire plants (Narboo) which causes mechanical wounding[104].

Seasonal incidence

Incidence of red spider mite, *T. urticae* is appered in the 14th standard week and reached to its maximum in the 12th standard week[106]. Similarly, in *kharif* season, the pest population strats to appear in presence of dry spell from the August and its peak population is noticed in the month of October and November[111].

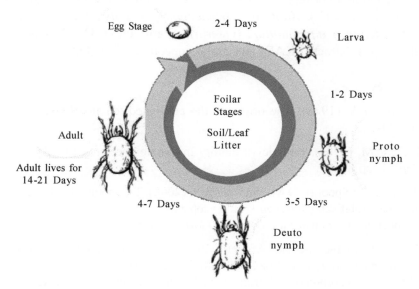

Fig. Diagrammatic representaton of life cycle of spider mite

Biology

Spider mites, Tetranichus urticae produces approximately 7 to 8 eggs per day either sexually or parthenogenetically which hatch out in 2-4 days[107]. Fecundity of mated females is more than unmated females. There are basically three immature stages in spider mite i.e. larva, protonymph and deutonymph. Protonymph is form after first moulting and last for 3-5 days and again in counter ecdysis to form deutonymph which remain for the period of 4-7 days[109]. An emergence of final adult stage from egg and all intermediate stages takes approximately 16-18 days[101].

It is observed that total duration from egg to adult emergence or the total developmental period was less in male as compared to female[105].

Biointensive Integrated/Insect Pest Management

Many notable natural enemies are found in environment which are effective against spider mites for example, predatory mites (Phytoseiulus persimilis) which is the natural enemy of *T. urticae* and is commonly used to control spider mites on tomato[102,103].

Entomopathogenic fungi recommended for the monitoring of mites are *Beauveria bassiana* and *Metarhizium anisoplae*[113].

The many plants and their extracts bearacaricidal properties which are being widely used in monitoring the population of spider mites infesting tomato such as*Acorus calamus* (L), *Xanthium strumarium* (L), *Polygonum hydropiper* (L), *Clerodendron infortunatum* (Gaertn), Azadirachtin 3% WSP,pongam kernel aqueous extract (PKAE) and garlic aqueous extract (GAE)[107,112].

References

1. Jenkins, J.A. (1948). The origin of the cultivated tomato. *Econ. Bot.*, 2, pp 379–392.

2. Rick, C.M. (1983). Genetic variability in tomato species. *Plant Mol. Biol. Rep.*, 1, pp 81–87.

3. Peralta, I.E.; Spooner, D.M. and Knapp, S. (2008). Taxonomy of wild tomatoes and their relatives (*Solanum* sect. Lycopersicoides, sect. Juglandifolia, sect. Lycopersicon; Solanaceae). *Syst. Bot. Mon.*, 84, pp 1-186.

4. José Luis Chávez-Servia; Araceli Minerva Vera-Guzmán;Lesser Roberto Linares-Menéndez; José Cruz Carrillo-Rodríguez and Elia Nora Aquino-Bolaños (2018). Agromorphological Traits and Mineral Content in Tomato Accessions from El Salvador, Central America. *Agronomy,* 8(32): 1-14. doi:10.3390

5. White, P.J. and Broadley, M.R. (2005). Biofortifying crops with essential mineral elements. *Trend Plant Sci.*, 10: 586-593.

6. Oyetayo, F.L. and Ibitoye, M.F. (2012). Phytochemical and nutrient/antinutrient interactions in cherry tomato (Lycopersicum esculentum) fruits. *Intern. J. Adv. Biol. Res.*, 2: 681-684.

7. Hartman, J.B. and St Clair, D.A. (1999). Combining ability for beet armyworm, *Spodoptera exigua*, resistance and horticultural traits of selected *Lycopersicon pennellii*-derived inbred backcross lines of tomato. *Plant Breed.* 118: 523-530.

8. Zhudong Liu; Dianmo Li; Peiyu Gong and Kunjun Wu (2004). Life Table Studies of the Cotton Bollworm, *Helicoverpa armigera* (Hübner) (Lepidoptera: Noctuidae), on Different Host Plants, *Environmental Entomology*, 33(6): 1570-1576.

9. Perkins, L.E.; Cribb, B.W.; Hanan, J.; Glaze, E.; Beveridge, C. and Zulucki, M.P. (2008). Development and feeding behavior of *Helicoverpa armigera* (Hübner) (Lepidoptera: Noctuidae) on different sunflower genotypes under laboratory conditions, Arthropod-Plant Interactions, Springer, 2: 197. https://doi.org/10.1007/s11829-008-9047.

10. Zalucki, M.P.; Murray, D.A.H.; Gregg, P.C.; Fitt, G.P.; Twine, P.H. and Jones, C. (1994). Ecology of *Helicoverpa armigera* (Hübner) and *H. punctigera* (Wallengren) in the inland of Australia: larval sampling and host plant relationships during winter and spring. *Australian Journal of Zoology*; 42: 329-346.

11. Fathipour, Y. and Naseri, B. (2011). Soybean Cultivars Affecting Performance of Helicoverpa armi gera (Lepidoptera: *Noctuidae*). In: Ng TB. (ed.) Soybean - Biochemistry, Chemistry and Physiology. Rijeka: InTech; Pp 599-630.

12. Reed, W. and Pawar, C.S. (1981). Heliothis: A Global Problem. In: Reed W, Kumble V. (eds.) Pro ceedings of the International Workshop on Heliothis Management, 15-20 November 1981, Patancheru, India. International Crops Research Institute for the Semi-Arid Tropics; 1982. p 9-14.

13. Bajpai, N.K. and Sehgal, V.K., (1994). Effect of neem products, nicotine and karanj on survival and biology of pod borer, *Helicoverpa armigera* Hubn. of chickpea. *Proc. II AZRA Conf. on Recent Trends in Plant, Animal and Human Pest Management: Impact on Environment.* Madras Christian College, Madras, 27-29 pp 48.

14. Sinha, S.H. (1993). Neem in the integrated management of *Helicoverpa armigera* Hubn. in chickpea. *World Neem Conference,* Bangalore, India 24-28 Feb. 1993. pp 6.

15. Chandra Shekhara; V. Rachappa; Suhas Yelshetty and A.G. Sreenivas (2014). Biorationals for eco-friendly management of gram pod borer, *Helicoverpa armigera* (HUBNER) on chickpea. *J. Exp. Zool. India.* 17(2): 679-682.

16. Dubey, N.K.; R. Shukla; A. Kumar; P. Singh and B. Prakash (2011). Global Scenario on the application of Natural products in integrated pest management. *In: Natural Products in Plant Pest Management,* Dubey, N.K., CABI, Oxfordshire, UK., pp: 1-20.

17. Khater, H.F., (2011). Ecosmart Biorational Insecticides: Alternative Insect Control Strategies. In: Insecticides, Perveen, F. (Ed.). InTech, Rijeka, Croatia, ISBN 979-953-307-667-5.

18. Anand Prakash and Jagadiswari Rao (2016). Biointensive Insect Pest Management,National Symposium on "New Horizons in Pest Management for Sustainable Goals" on 24-25 November, 2016 at OUAT, Bhubaneswar.

19. Fathipour, Y. and Sedaratian, A. (2013). Integrated Management of *Helicoverpa armigera* in Soybean Cropping Systems. IntechOpen. DOI: 10.5772/54522

20. Sapkal, S.D.; Sonkamble, M.M. and Gaikwad, B.B. (2018). Seasonal incidence of tomato fruit borer, *Helicoverpa armigera* (Hubner) on tomato, *Lycopersicon esculentum* (Mill) under protected cultivation. *Journal of Entomology and Zoology Studies* 2018; 6(4): 1911-1914.

21. Harshita, A.P.; Saikia, D.K.; Anjumoni Deeve; Bora, L.C. and Phukan, S.N. (2018). Seasonal incidence of fruit borer, *Helicoverpa armigera* and its eco-friendly management in tomato, *Solanum lycopersicum. International Journal of Chemical Studies*; 6(5): 1482-1485.

22. Gomes, E.S.; Santos, V. and Ávila, C.J. (2017). Biology and fertility life table of *Helicoverpa armigera* (Lepidoptera: *Noctuidae*) in different hosts. *Entomological Science*, 20: 419–426.

23. Patil, V.M.; Patel, Z.P.; Oak, P.S.; Chauhan, R.C. and Kaneriya, P.B. (2018). Biology of fruit borer, *Helicoverpa armigera* (Hubner) in/on chilli fruits. *International Journal of Entomology Research*, 3(1): 6-12.

24. Munir Ahmad; Abdul Ghaffar and Muhammad Rafiq (2013). Host Plants of Leaf Worm, *Spodoptera Litura* (Fabricius) (Lepidoptera: *Noctuidae*) in Pakistan. *Asian J Agri Biol*, 1(1): 23-28.

25. Xue, M.; Pang, Y.H.; Wang, H.T.; Li, Q.L. and Liu, T.X. (2010). Effects of four host plants on biology and food utilization of the cutworm, *Spodoptera litura. Journal of Insect Science* 10(22): 1-14.

26. Fand, B.B.; Sul, N.T.; Bal, S.K. and Minhas, P.S. (2015). Temperature Impacts the Development and Survival of Common Cutworm (*Spodoptera litura*): Simulation and Visualization of Potential Population Growth in India under

Warmer Temperatures through Life Cycle Modelling and Spatial Mapping. PLoS ONE 10(4): e0124682.

27. IIE (1993). Distribution Maps of Plant Pests, Wallingford (UK): CAB International. No. 61.

28. EPPO (2013). PQR database: European and Mediterranean Plant Protection Organization, Paris, France. Available: http://www.eppo.int/DATABASES/pqr/pqr.htm.

29. Moussa, A.M.; Zather, M.A. and Kothy, F. (1960). Abundance of cotton leaf worm, *Prodenia litura* (F) (Lepidoptera: Agrotidae-Zanobiinae) in relation to host plants and their effects on biology. *Bulle. Soc. Entomol. Egypt.* 44: 241-251.

30. CABI (2014). Invasive Species Compendium: Datasheets, maps, images, abstracts and full text on invasive species of the world. Available: http://www.cabi.org/isc/datasheet/44520.

31. Garad, G.P.; Shivapuje, P.R. and Bilapate, G.G. (1984). Life fecundity tables of *Spodoptera litura* (F.) on different hosts. Proc. *Indian Acad. Sci.,* 93: 29-33.

32. Choudhary, A.K. and Srivastava, S.K. (2007). Efficacy and economics of some neem based products against tobacco caterpillar, *Spodoptera litura* F. on soybean in Madhya Pradesh, *India. Int. J. Agric. Sci.,* 3: 15-17.

33. Punithavalli, M.; Sharma, A.N. and Balaji, R.M. (2014). Seasonality of the common cutworm *Spodoptera litura* in a soybean ecosystem. *Phytoparasitica,* 42: 213-222.

34. Sharma, A.N.; Gupta, G.K.; Verma, R.K.; Sharma, O.P.; Bhagat, S. and Amaresan, N. (2014). Integrated Pest Management Package for Soybean. Faridabad (India): Directorate of Plant Protection, Quarantine and Storage. pp 41.

35. Narvekar, P.F.; Mehendale, S.K.; Golvankar, G.M.; Karmarkar, M.S. and Desai, S.D. (2018). Comparative biology of *Spodoptera litura* (Fab.) on different host plants under laboratory condition. *International Journal of Chemical Studies,* 6(6): 65-69.

36. Sushil Kumar and Puja Ray (2007). Biology of *Spodoptera litura* (Fab.) on some of its major weed hosts. *Entomon,* 34(4): 287-290.

37. Naik, C.M.; Nataraj, K. and Santhoshakumara, G.T. (2017). Comparative Biology of *Spodoptera litura* on Vegetable and Grain Soybean [*Glycine max* (L.) Merrill]. *Int. J. Curr. Microbiol. App. Sci.,* 6(7): 366-371.

38. Dhir, B.C.; Mohapatra, H.K. and Senapati, B. (1992). Assessment of crop loss in groundnut due to tobacco caterpillar, *Spodoptera litura* (F.). *Ind. J. Plant Prot.* 20: 215-217.

39. Vashisth, S.; Chandel, Y.S. and Kumar, S. (2012). Biology and damage potential of *Spodoptera litura* Fabricius on some important greenhouse crops. *Journal of Insect Science*, 25(2): 150-154.

40. Bale, J.; Masters, G.J.; Hodkins, I.D.; Awmack, C., Bezemer, T.M. and Brown, V.K. (2002). Herbivory in global climate change research: direct effects of rising temperature on insect herbivores. *Global Change Biol* 8: 1-16.

41. Gadad, H.; Hegde, M. and Balikai, R.A. (2013). Seasonal incidence of *Spodoptera litura* and leafminer in rabi/summer groundnut. *J. Exp. Zool. India* 16(2): 619-622.

42. Misal, M.R.; Patil, B.V. and Khaire, P.B. (2018). Management of *Spodoptera litura* in Soybean through Biorational Approaches. *Int.J. Curr. Microbiol. App.* Sci., 7(12): 216-222.

43. Nagal, G.; Verma, K.S. and Rathore, L. (2016). Management of *Spodoptera litura* (Fabricius) through some novel insecticides and biopesticides on bell pepper under polyhouse environment. *Journnal Advances in Life Sciences* 5(3):1081-1084.

44. Hara, A.H.; Kaya, H.K.; Gaugler, R.; Lebeck, L.M. and Mello, C.L. (1993). Entomopathogenic nematodes for biological control of the leaf miner, *Liriomyza trifolii* (Dipt.: Agromyzidae). *Entomophaga* 38: 359-369.

45. Leibee, G.L. (1984). Influence of temperature on development and fecundity of *Liriomyza trifolii* (Burgess) (Diptera: *Agromyzidae*) on celery. *Environmental Entomology*, 13: 497-501.

46. Levins, R.A.; Poe, S.L.; Littell, R.C. and Jones, J.P. (1975). Effectiveness of a leafminer control program for Florida tomato production. *Journal of Economic Entomology*, 68: 772-774.

47. Mason, G.A.; Tabashnik, B.E. and Johnson, M.W. (1989). Effects of biological and operational factors on evolution of insecticide resistance in *Liriomyza* (Diptera: Agromyzidae). *Journal of Economic Entomology*, 82: 369-373.

48. Minkenberg, O.P.J.M. (1988). Life history of the agromyzid fly *Liriomyza trifolii* on tomato at different temperatures. *Entomologica Experimentalis et Appliciata*, 48: 73-84.

49. Minkenberg, O.P.J.M. and Lenteren, J.C. (1986). The leafminers *Liriomyza bryoniae* and *L. trifolii* (Diptera: Agromyzidae), their parasites and host plants: a review. *Wageningen Agric. Univ. Papers*, 86-2.50 pp.

50. Murphy, S.T.; LaSalle, J. (1999). Balancing biological control strategies in the IPM of New World invasive *Liriomyza* leafminers in field vegetable crops. *Biocontrol News and Information*, 20: 91-104N.

51. Parrella, M.P.; Robb, K.L. and Bethke, J. (1983). Influence of selected host plants on the biology of *Liriomyza trifolii* (Diptera: *Agromyzidae*). *Annals of the Entomological Society of America*, 76: 112-115.

52. Parrella, M.P.; Jones, V.P.; Youngman, R.R. and Lebeck, L.M. (1985). Effect of leaf mining and leaf stippling of *Liriomyza* spp. on photosynthetic rates of chrysanthemum. *Annals of the Entomological Society of America*, 78: 90-93.

53. Parrella, M.P.; Yost, J.T.; Heinz, K.M. and Ferrentino, G.W. (1989). Mass rearing of *Diglyphus begini* (Hymenoptera: *Eulophidae*) for biological control of *Liriomyza trifolii* (Diptera: *Agromyzidae*). *Journal of Economic Entomology*, 82: 420-425.

54. Wolfenbarger, D.A. and Wolfenbarger, D.O. (1966). Tomato yields and leaf miner infestations and a sequential sampling plan for determining need for control treatments. *Journal of Economic Entomology*, 59: 279-283.

55. Zehnder, G.W. and Trumble, J.T. (1984). Spatial and diet activity of *Liriomyza* species (Diptera: *Agromyzidae*) in fresh market tomatoes. *Environmental Entomology*, 13: 1411-1416.

56. Parish, J.B.; Carvalho,G.A.; Ramos, R.S.; Queiroz, E.A.; Picanço, M.C.; Guedes, R.N.C. and Corrêa, R.C. (2017). Host range and genetic strains of leafminer flies (Diptera: *Agromyzidae*) in eastern Brazil reveal a new divergent clade of *Liriomyza sativae*. *Agricultural and Forest Entomology* 19(3): 235-244.

57. Spencer, K.A. (1990). Host Specialization in the World *Agromyzidae (Diptera)*. Kluwer Academic Publishers, The Netherlands.

58. Jayakumar, P. and Uthamasamy, S. (2000). Distribution and biology of *Liriomyza trifolii* (Burges) in Tamil Nadu, India. *International Journal of Tropical Insect Science*, 20(4): 237-243.

59. Selvaraj, S.; Bisht, R.S. and Ganeshamoorthi, P. (2016). Seasonal incidence of American serpentine leaf miner, *Liriomyza trifolii* (BURGESS), on Tomato at Pantnagar, Uttarakhand. *International Journal of Agriculture Sciences*, 8(38): 1777-1779.

60. Chakraborty, K. (2011). Incidence and Abundance of Tomato Leaf Miner, *Liriomyza trifolii* Burgess in Relation to the Climatic Conditions of Alipurduar, Jalpaiguri,West Bengal, *India. Asian J. Exp. Biol. Sci.* 2(3): 467-473.

61. Shilpakala, V. and Krishna, T.M. (2016). Seasonal Incidence of Serpentine Leaf Miner, *Liriomyza Trifolii* and influence of weather parameters on its incidence in different sowings of castor cultivars. *International Journal of Science, Environment and Technology*, 5(6): 3742-3748.

62. Ramesh, R. and Ukey, S.P. (2007). Bio-efficacy of botanicals, microbials and newer insecticides in the management of tomato leafminer, *Liriomyza trifolii* Burgess. *Intl. J. Agri.Sci.* 3(1): 154-156.

63. Suradkar, A. and Ukey, S.P. (2014). Bio-efficacy of botanicals, microbial and conventional insecticide against tomato leaf miner. *Agriculture for Sustainable Development*, 2(2): 92-96.

64. Hara, A.H.; Kaya, H.K.; Gaugler, R.; Lebeck, L.M. and Mello, C.L. (1993). Entomopathogenic nematodes for biological control of the leafminer, *Liriomyza trifolii* (Dipt.: *Agromyzidae*) *Entomophaga* 38:359. https://doi.org/10.1007/BF02374454.

65. Liu, T.X.; Kang, L.; Heinz, K.M. and Trumble, J. (2009). Biological control of *Liriomyza* leafminers: progress and perspective. CAB Reviews: *Perspectives in Agriculture, Veterinary Science, Nutrition and Natural Resources*, 4: 004.

66. Kumar, P.; Naqvi, A.R.; Meena, R.S. and Mahendra M. (2019). Seasonal incidence of whitefly, *Bemisia tabaci* (Gennadius) in tomato (*Solanum lycopersicum* Mill). *International Journal of Chemical Studies*, 7(3): 185-188.

67. Patra, B.; Alam, S.F. and Samanta, A. (2016). Influence of Weather Factors on the Incidence of Whitefly, *Bemisia tabaci* Genn. on Tomato in Darjeeling Hills of West Bengal. *Journal of Agroecology and Natural Resource Management,* 3(3): 243-246.

68. Sharma, D.; Maqbool, A.; Jamwal, V.V.S.; Srivastava, K. & Sharma, A. (2017). Seasonal dynamics and management of whitefly (*Bemesia tabaci* Genn.) in tomato (*Solanum esculentum* Mill.).Brazilian Archives of Biology and Technology, 60, e17160456.Epub.https://dx.doi.org/10.1590/1678-4324-2017160456.

69. Jha, S.K. and Kumar, M. (2018). Fluctuation in whitefly *Bemisia tabaci* population in relation to environmental factors. *Journal of Entomology and Zoology Studies*, 6(2): 3011-3014.

70. Atwal, A.S. and Dhaliwal, G.S. (2005). Agricultural pests of South Asia and Their Management. Kalyani Publishers, Ludhiana, 216.

71. Arnal, E.; Debrot, E.; Marcano, R.M. and Manlagne, A. (1998). Population fluctuation of whitefly and its relation to tomato yellow mosaic in one location in Venezuela. *Fitopatologia Venezoloana*. 6(1): 21-26.

72. Ghosh, S.K.; Laskar, N.; Basak, S.N. and Senapati, S.K. (2004). Seasonal fluctuation of *Bemisia tabaci* on Brinjal and field evaluation of some pesticide against *Bemisia tabaci* under Tarai region of West Bengal. *Environment and Ecology*, 22(4): 758-762.

73. CABI/EPPO, (2014). Bemisia tabaci.[Distribution map]. Distribution Maps of Plant Pests, June (1st revision). Wallingford, UK: CAB International, Map 284.

74. Cuthbertson, A.G.S.; Buxton, J.H.; Blackburn, L.F.; Mathers, J.J.; Robinson, K.A.; Powell, M.E.; Fleming, D.A. and Bell, H.A. (2012). Eradicating *Bemisia tabaci* Q biotype on poinsettia plants in the UK. *Crop Protection*, 42: 42-48.

75. Evans, G.A. and Polaszek, A. (1997). Additions to the *Encarsia* parasitoids (Hymenoptera: *Aphelinidae*) of the *Bemisia tabaci*-complex (Hemiptera: *Aleyrodidae*). *Bulletin of Entomological Research,* 87(6): 563-571.

76. Eslamizadeh, R.; Sajap, A.S.; Omar, D. and Adam, N.A. (2013). First record of *Isaria fumosorosea* Wize (Deuteromycotina: *Hyphomycetes*) infecting *Bemisia tabaci* (Gennadius) (Hemiptera: *Aleyrodidae*) in Malaysia. *Journal of Entomology*, 10(4): 182-190.

77. Cuthbertson, A.G.S.; Head, J.; Walters, K.F.A. and Gregory, S. A. (2003). The efficacy of the entomopathogenic nematode, *Steinernema feltiae*, against the immature stages of *Bemisia tabaci*. *Journal of Invertebrate Pathology*, 83: 267–269.

78. Cuthbertson, A.G.S.; Mathers, J.J.; Northing, P.; Qi, L.W. and Walters, K.F.A. (2007). The susceptibility of immature stages of Bemisia tabaci to infection by the entomopathogenic nematode *Steinernema carpocapsae*. *Russian Journal of Nematology*, 15(2): 153-156.

79. Cuthbertson, A.G.S. and Walters, K.F.A. (2005). Pathogenicity of the entomopathogenic fungus, *Lecanicillium muscarium*, against the sweet potato whitefly *Bemisia tabaci*, under laboratory and glasshouse conditions. *Mycopathologia*, 160(4): 315-319.

80. Cuthbertson, A.G.S. (2013). Update on the status of *Bemisia tabaci* in the UK and the use of entomopathogenic fungi within eradication programmes. *Insects*, 4(2): 198-205.

81. Cuthbertson, A.G.S. (2014). The feeding rate of predatory mites on life stages of *Bemisia tabaci* Mediterranean species. *Insects*, 5(3): 609-614.

82. Khan, I.A. and Wan, F.H. (2015). Life history of *Bemisia tabaci* (Gennadius) (Homoptera: *Aleyrodidae*) biotype B on tomato and cotton host plants. *Journal of Entomology and Zoology Studies*, 3(3): 117-121.

83. Gangwar, R.K. and Gangwar, C. (2018). Lifecycle, Distribution, Nature of Damage and Economic Importance of Whitefly, *Bemisia tabaci* (Gennadius). *Acta Scientific Agriculture* 2(4): 36-39.

84. Kedar, S.C.; Saini, R.K. and Kumaranag, K.M. (2014). Biology of cotton whitefly, *Bemisi tabaci* (Hemiptera: *Aleyrodidae*) on cotton. *J. Ent. Res.*, 38(2): 135-139.

85. Fekrat, L. and Shishehbor, P. (2007). Some Biological Features of Cotton Whitefly, *Bemisia tabaci* (Homoptera: *Aleyrodidae*) on Various Host Plants. *Pakistan Journal of Biological Sciences*, 10(18): 3180-3184.

86. Nzanza, B. and Mashela, P.W. (2012). Control of whiteflies and aphids in tomato (*Solanum lycopersicum* L.) by fermented plant extracts of neem leaf and wild garlic. *African Journal Of Biotechnology*, 11(94): 16077-16082.

87. Jenser, G.; Lipcsei, S.; Szénási, Á. and Hudák, K. (2006). Host Range of the Arrhenotokous Populations of *Thrips Tabaci* (Thysanoptera: *Thripidae*). *Acta Phytopathologica et Entomologica Hungarica*, 41(3-4): 297-303.

88. Van Rijn, P.; Mollema, C. and Steenhuis-Broers, G. (1995). Comparative life history studies of *Frankliniella occidentalis* and *Thrips tabaci* (Thysanoptera: Thripidae) on cucumber. *Bulletin of Entomological Research*, 85(2): 285-297.

89. Sakimura, K. (1932). Life History of *Thrips Tabaci* L. on *Emilia Sagittata* and Its Host Plant Range in Hawaii. *Journal of Economic Entomology*, 25(4): 884-891.

90. Ananthakrishnan, T.N. (1993). Bionomics of Thrips. *Annual Review of Entomology*, 38(1): 71-92.

91. Morse, J.G. and Hoddle, M.S. (2006). Invasion Biology of Thrips. *Annual Review of Entomology*, 51: 67-89.

92. Jenser, G. and Szénási Á. (2004). Review of the biology and vector capability of Thrips tabaci Lindeman (Thysanoptera: Thripidae). *Acta Phytopathologica et Entomologica Hungarica,* 39(1-3): 14.

93. Mark Hoddle (2011). Western Flower Thrips in Greenhouses: A Review of its Biological Control and Other Methods.Applied Biological Control Research, https://biocontrol.ucr.edu/wft.html#Summary.

94. Gawaad, A.A.A. and Soliman, A.S. (2009). Studies on *Thrips tabaci* Lindman. *Journal of Applied Entomology* 70(1-4): 93-98.

95. Radhika, P. (2013). Influence of weather on the seasonal incidence of insect pests on groundnut in the scarce rainfall zone of Andhra Pradesh. *Adv. Res. J. Crop Improv.*, 4(2): 123-126.

96. Zainab, S.; Sathua, S.K. and Singh, R.N. (2016). Study of population dynamics and impact of abiotic factors on thrips, *Scirtothrips dorsalis* of chilli, *Capsicum annuum* and comparative bio-efficacy of few novel pesticides against it. *International Journal of Agriculture, Environment and Biotechnology*, 9(3): 451-456.

97. Mandloi, R.; Pachori, R.; Sharma, A.K.; Thomas, M.andThakur, A.S. (2015). Impact of weather factors on the incidence of major insect pests of Tomato (*Solanum lycopersicon* L.) CV. H-86 (Kashi Vishesh). *The Ecoscan*, 7: 7-15.

98. Gill, H.K.; Garg, H.; Gill, A.K.; Gillett-Kaufman, J.L. and Nault, B.A. (2015). Onion Thrips (Thysanoptera: *Thripidae*) Biology, Ecology, and Management in Onion Production Systems. *Journal of Integrated Pest Management*, 6(1).

99. Boateng, C.O.; Schwartz, H.F.; Havey, M.J. and Otto, K. (2014). Evaluation of onion germplasm for resistance to Iris Yellow Spot Virus (Iris yellow spot virus) and onion thrips, *Thrips tabaci*. Southwest. *Entomol.*, 39: 237-260.

100. Shiberu, T.; Negeri, M. and Selvaraj, T. (2013). Evaluation of Some Botanicals and Entomopathogenic Fungi for the Control of Onion Thrips (*Thrips tabaci* L.) in West Showa, Ethiopia. *J. Plant Pathol. Microb.*, 4: 161.

101. Kant, M.R.; Ament, K.; Sabelis, M.W.; Haring, M.A. and Schuurink, R.C. (2004). Differential Timing of Spider Mite-Induced Direct and Indirect Defenses in Tomato Plants. *Plant Physiology*, 135: 483-495.

102. Drukker, B.; Janssen, A.; Ravensberg, W. and Sabelis, M.W. (1997). Improved control capacity of the mite predator *Phytoseiulus persimilis* (Acari: Phytoseiidae) on tomato. *Exp Appl Acarol* 21: 507-518.

103. Garthwaite, D. (2000). Changes in biological control usage in Great Britain between 1968 and 1995 with particular reference to biological control on tomato crops. *Biocontrol Sci Techn* 10: 451-457.

104. Strassner, J.; Schaller, F.; Frick, U.B.; Howe, G.A.; Weiler, E.W.; Amrhein, N.; Macheroux, P. and Schaller, A. (2002). Characterization and cDNA-microarray expression analysis of 12-oxophytodienoate reductases reveals differential roles for octadecanoid biosynthesis in the local versus the systemic wound response. *Plant J.* 32: 585-601.

105. Satish, S.B.; Pradeep, S.; Sridhara, S.; Narayanaswamy, H. and Manjunatha, M. (2018). Biology of Red Spider Mite, *Tetranychus macfarlanei* Baker and Pritchard on Soybean. *International Journal of Microbiology Research*, 10(9): 1370-1373.

106. Norboo, T.; Ahmad, H; Shankar, U.; Ganai, S.A.; Khaliq, N. and Mondal, A. (2017). Seasonal Incidence and Management of Red Spider Mite, *Tetranychus urticae* Koch. infesting Rose. *Int.J. Curr. Microbiol. App. Sci.,* 6(9): 2723-2729.

107. Takafuji, A.; Ozawa, A.; Nemoto, H. and Gotoh, T. (2000). Spider Mites of Japan: Their Biology and Control. *Experimental & Applied Acarology*, 24(5-6): 319-335.

108. Navajas, M. (1998). Host plant associations in the spider mite *Tetranychus urticae* (Acari: *Tetranychidae*): insights from molecular phylogeography. *Experimental & Applied Acarology*, 22(4): 201-214.

109. Boudreaux, H.B. (1963). Biological aspects of some phytophagous mites. *Annu. Rev. Entomol.*, 8: 137-154.

110. Gotoh, T.; Bruin, J.; Sabelis, M.W. and Menken, S.B.J. (1993). Host race formation in *Tetranychus urticae*: genetic differentiation, host plant preference, and mate choice in a tomato and a cucumber strain. *Entomol. Exp. Appl.* 68: 171-178.

111. Migeon, A.; Ferragut, F.; Escudero-Colomar, L.A.; Fiaboe, K.; Knapp, M.; Moraes, G.; Ueckermann, E. and Navajas, M. (2009). Modelling the potential distribution of the invasive tomato red spider mite, *Tetranychus evansi* (Acari: *Tetranychidae*). *Experimental and Applied Acarology*, 48(3): 199-212.

112. Sarmah, M.; Rahman, A.; Phukan, A.K. and Gurusubramanian, G. (2009). Effect of aqueous plant extracts on tea red spider mite, *Oligonychus coffeae*, Nietner (*Tetranychidae:* Acarina) and *Stethorus gilvifrons* Mulsant. *African Journal of Biotechnology*, 8(3): 417-423.

113. Amjad, M.; Bashir, M.H.; Afzal, M.; Sabri, M.A. and Javed, N. (2012). Synergistic Effect of Some Entomopathogenic Fungi and Synthetic Pesticides, Against Two Spotted Spider Mite, *Tetranychus urticae* Koch (Acari: *Tetranychidae*). *Pakistan J. Zool.*, 44 (4): 977-984.

14

Biointensive Integrated Pest Management of Potato

Sagar Tamang, Tanweer Alam and Lalita Rana

Introduction

Potato (*Solanum tuberosum*) is the most important food crop of the world. Potato is a temperate crop grown under subtropical conditions in India. The potato is a crop which has always been the 'poor man's friend'. Potato is being cultivated in the country for the last more than 300 years. For vegetable purposes it has become one of the most popular crops in this country. Potatoes are an economical food; they provide a source of low cost energy to the human diet. Potatoes are a rich source of starch, vitamins especially C and B1 and minerals. They contain 20.6 per cent carbohydrates, 2.1 per cent protein, 0.3 per cent fat, 1.1 per cent crude fibre and 0.9 per cent ash. They also contain a good amount of essential amino acids like leucine, tryptophane and isoleucine etc.

Potatoes are used for several industrial purposes such as for the production of starch and alcohol. Potato starch (farina) is used in laundries and for sizing yarn in textile mills. Potatoes are also used for the production of dextrin and glucose. As a food product itself, potatoes are converted into dried products such as 'potato chips', 'sliced' or 'shredded potatoes'.

Potato is grown almost in all states of India. However, the major potato rowing states are Himachal Pradesh, Punjab, Uttar Pradesh, Madhya Pradesh, Gujarat, Maharashtra, Karnataka, West Bengal, Bihar and Assam.

The main constraint to potato farming in India are: It is vulnerable to control the pests hence implying a high risk of failure, growing potatoes requires substantial capital and the crop needs intensive care and attention. However, a proper Biointensive insect pest management programme will minimize losses to potato crop.

1. Potato tuber moth: *Phthorimaea operculella*

Taxonomy

Kingdom	:	Metazoa
Phylum	:	Arthropoda
Subphylum	:	Uniramia
Class	:	Insecta
Order	:	Lepidoptera
Family	:	Gelechiidae
Genus	:	*Phthorimaea*
Species	:	*operculella*

Symptoms of damage

- The tuber moth is a pest of field and storage
- Larva tunnels into foliage, stem and tubers
- Galleries are formed near tuber eyes
- On growing plants, leaf mines betray the presence of larvae, and in addition the stem is weakened or broken. On tubers, detection is more difficult without cutting open some tubers, when galleries and larvae will be apparent within the potatoes

Biology

Egg

Laid singly on the ventral surface of foliage and exposed tubers

Larva

When fully grown, *P. operculella* larvae are about 15 mm in length. Head dark brown; prothoracic plate sometimes pinkish; body greyish-white or pale greenish-grey. Pinacula small, dark brown or black. Anal plate brown.

Pupa

Pupation occurs within a cocoon among the trash, clods of the earth in the field. Pupation normally takes place in the soil; the pupal period is 6-29 days.

Adult

Small narrow winged moth, measuring about 1 cm in length when at rest, greyish brown forewings and hind wings are dirty white. Adults fly chiefly by night and are attracted to light. They live for up to 10 days. The moth breeds continuously where conditions permit; up to 13 generations a year have been recorded in India (Mukherjee, 1948). The complete life cycle ranges from 17 to 125 days.

Seasonal Incidence

2. Cut worm: *Agrotis ipsilon*

Taxonomy

Kingdom	:	Animalia
Phylum	:	Arthropoda
Subphylum	:	Uniramia
Class	:	Insecta
Order	:	Lepidoptera
Family	:	Noctuidae
Genus	:	*Agrotis*
Species	:	*ipsilon*

Symptoms of damage

- Young larvae feed on the epidermis of the leaves.
- Older larvae come out at night and feed young plants by cutting their stems.
- They also damage the tubers by eating away part of them.

Biology

Eggs

The egg stage lasts 3 to 6 days. Females oviposit eggs in clusters on low lying leaves. If such host plants are not available, the females will oviposit on dead plant material. However, they will not lay eggs on bare soil. Females can deposit eggs singly, or in groups of up to 1200 to 1900 eggs.

Caterpillar

The larval stage lasts 20-40 days. Over the span of 5 to 9 instars, the caterpillar body grows from 3.5 mm to a maximum of 55 mm. By the 4th instar, the larva becomes light sensitive and spends most of the daylight underground. The larvae are cannibalistic in nature. The full-grown larva is dark or dark brown with a plump and greasy body.

Pupa

The pupal stage lasts 12-20 days. This species pupates under the soil approximately 3-12 mm below the surface. The pupae appear to be dark brown and are 17-12 mm long and 5-6 mm wide.

Adult

One complete generation from egg to adult lasts 35-60 days. The female preoviposition period lasts 7-10 days. Adults have a wingspan of 40-55 mm. The forewings are dark brown, and the distal area has a light irregular band a black dash mark.

Seasonal Incidence

Persistent dry weather with lesser or no rainfall, reduced humidity & 16-23°C temperature favor the development of cutworm.

3. Green peach Aphid: *Myzus persicae*

Taxonomy

Kingdom	:	Animalia
Phylum	:	Arthropoda
Class	:	Insecta
Order	:	Hemiptera
Family	:	Aphididae
Genus	:	*Myzus*
Species	:	*persicae*

Damage symptoms

Direct damage

- Aphids damage plants by puncturing them and sucking their juices.
- They damage the young and soft parts of plants, such as new leaves and shoots.
- Signs of damage are leaves not opening properly and being smaller in size
- Severe infestation can cause shoots to wilt and dry out.

Indirect damage

- Aphids have wings and can move from plant to plant spreading viral diseases, picked up from infected plants.

- Aphids secrete a sugary liquid that stimulates black sooty mold growth. It can cover the surface of leaves which affects the way they absorb sunlight

Biology

Aphids reproduce in two ways: by laying eggs and giving birth to young ones. Which birth process is used depends on environmental conditions and the availability of food.

Eggs

Eggs are deposited on *Prunus spp.* trees. The eggs measure about 0.6 mm long and 0.3 mm wide, and are elliptical in shape. Eggs initially are yellow or green, but soon turn black. Mortality in the egg stage sometimes is quite high. When food is plentiful, aphids give birth to live young.

Nymphs

Nymphs initially are greenish, but soon turn yellowish, greatly resembling viviparous (parthenogenetic, nymph-producing) adults. Horsfall (1924) He reported that four instars in this aphid, with the duration of each averaging 2.0, 2.1, 2.3, and 2.0 days, respectively.

Adults

Eggs hatch after three or four days. Young aphids, called nymphs, need five to eight days to become adults. Up to 8 generations may occur on Prunus in the spring, but as aphid densities increase winged forms are produced, which then disperse to summer hosts.

Seasonal Incidence

- In December, winged germs appear on potato crop. Apterous females are formed on potato, till end of March.
- With rise of temperature, winged forms are produced which migrate to hill areas again to primary hosts.
- Thus this pest appears in plain areas in winter while in the hills it appears in summer and autumn season. Its population sharply decreases with increase in RH over 73 percent.
- A relative humidity of 66 + 2.8% and 11-14°C temperature are ideal for development of this aphid.

4. Whitefly: *Bemisia tabaci*

Taxonomy

Kingdom	:	Animalia
Phylum	:	Arthropoda
Class	:	Insecta
Order	:	Hemiptera
Suborder	:	Sternorrhyncha
Superfamily	:	Aleyrodoidea
Family	:	Aleyrodidae
Genus	:	*Bemisia*
Species	:	*tabaci*

Damage symptoms

- Chlorotic spots
- Yellowing
- Downward curling and drying of leaves.
- Vector of potato leaf curl disease

Biology

Egg

The female whiteflies lay eggs singly on the underside of the leaves. Eggs are smooth, sub elliptical, stalked, and broader at basal end. Its colour is light yellow, when freshly laid, turn dark brown later on. The eggs hatch in 5-17 days. Stalked nymph: It is louse like, sluggish creature having pale-yellow body. The nymphal stage lasts 14 to 81 days.

Pupa

Convex in shape and possesses deep yellow patches on the abdomen.

Adult

In 2-8 days, the pupae change into white flies. Wings are pure white and have prominent long legs. The life cycle is completed in 14-122 days. Eleven generations of this pest are completed in a year.

Seasonal Incidence

The temperature of 28-36°C and 62-92% relative humidity and scanty rainfall during August to January are quite favorable for this pest.

5. Thrips: (*Thrips tabaci*)

Taxonomy

Kingdom	:	Animalia
Phylum	:	Arthropoda
Class	:	Insecta
Order	:	Thysanoptera
Family	:	Thripidae
Subfamily	:	Thripinae
Genus	:	*Thrips*
Species	:	*tabaci*

Damage

- In severely damaged plants, leaves may wither and the whole plant may appear silvery; the crop ripens prematurely but the yield is greatly reduced.
- Likely to find feeding damage along the veins, and on the back of the leaf
- May find damage on both sides of leaves but does not show through from one side to the other
- Feeding damage silvery in colour and shiny
- As thrips feed, the cells of the plant collapse forming pits, distortions and brown patches on the leaves. later, the leaves have a silver sheen, or a "scaring" occurs on the fruits.
- Heavy infestations cause leaves to wilt, dry and die

Biology

Thrips have a complex life cycle in which the last two immature stages are hidden and non-feeding. The life cycle has six stages: egg, first and second larva (small, whitish and very active, found mainly on the underside of the leaf), pre-pupa, pupa (both stages occur in the soil in natural cracks), and adult.

Egg

Thrips eggs are laid inside plant tissue. The female inserts her saw like ovipositor into plant tissues and lays her eggs under the epidermis, females lay about eighty eggs. The eggs are white at first, turning orange later, and hatch in four to five days.

Nymph

Two larval stages lasting about nine days in total are followed by the non-feeding prepupal and pupal stages which last four to seven days in total. The larvae are white or yellowish and suck sap from the plant tissues.

Pupa

Moves, feed, and show the developing wings.

Adult

The adult survives for two or three weeks during which time the females lay about eighty eggs. Most of the eggs are unfertilised and produced by parthenogenesis.

6. Leaf minor: (*Liriomyza* spp.)

Taxonomy

Kingdom	:	Animalia
Phylum	:	Arthropoda
Class	:	Insecta
Order	:	Diptera
Family	:	Agromyzidae
Subfamily	:	Phytomyzinae
Genus	:	*Liriomyza*
Species	:	*Liriomyza* spp.

Damage symptoms

- The leaf miner flies damage plants during its larval and adult stages mainly on the lower third of plants.
- Larvae begin eating the insides of leaves immediately after hatching, and bore mines inside them.
- In instances of severe infestation, all that is left of leaves is their upper and lower skins. Affected leaves become dry and drop off the plant.

- Adult flies puncture holes in leaves in order to lay eggs and feed on plant juices.

Biology

It was introduced into India through Chrysanthemum cuttings.

Egg

Adult flies produce an average of 166 eggs per female. Eggs are laid inside leaves and the eggs were hatches in 4 days., they are very small and clear in color. Larvae hatch after about 2-3 days.

Larva

Larvae remain inside leaves. They are very small and have no legs so cannot move from one leaf to another. The larval stage lasts around 6-12 days. Full grown larvae measure 3 mm

Pupa

These are formed in the ground or inside leaves. On potato plants, pupae usually fall to the ground. The pupal stage lasts around 14-16 days. Pupation takes place inside a thin loose mesh of silken cocoon.

Adult

These are extremely small at 2-4 mm in length, black in color with two yellow spots on their backs, measuring 1.5 mm in length. They are most active in the morning from 7:00 to 9:00 and in the afternoon from 16:00 to 18:00. They are attracted to the color yellow. Total life cycle takes 3 weeks.

7. Jassid: *Amrasca biguttula* (Ishida)

Jassid is the most destructive pest of many plants including Okra, brinjal, potato and cotton etc. It causes damage to potato in autumn season.

Taxonomy

Phylum	:	Arthropoda
Class	:	Insecta
Order	:	Hemiptera
Suborder	:	Auchenorrhyncha
Infraorder	:	Cicadomorpha

Superfamily	:	Membracoidea
Family	:	Cicadellidae
Subfamily	:	Typhlocybinae
Tribe	:	Empoascini
Genus	:	*Amrasca*
Species	:	*Biguttula*

Damage symptoms

- Heavy infestations on cotton, okra and sunflower make the leaves turn yellow, curl up and fall off.

- The insects also secrete honeydew, and sooty mould often grows on this, restricting the amount of light reaching the plant's photosynthetic surfaces and reducing the yield.

- In many areas, this pest regularly occurs on cotton in epidemic numbers.

- A number of natural enemies help to control populations including ladybirds, predatory lygaeid beetles, and several species of mantis. Neem oil can be used as a biopesticide.

Biology

- Leafhoppers undergo direct development from nymph to adult without undergoing metamorphosis.

- On okra, eggs are mainly oviposited inside the tissue of leaf blades, but may also be laid in leaf stalks or in soft twigs.

- The eggs hatch in six or seven days. There are five nymphal instars, developing over a period of about seven days.

- Total lifespan is about one month, with females living a little longer than males. The fecundity of females is about fifteen eggs.

Seasonal Incidence

- The adult Jassid breeds practically throughout the year, but during the winter months only adults are found on potato, brinjal and tomato, etc.

8. Hadda / spotted beetle: *H.vigintioito punctata*
Taxonomy

Kingdom	:	Animalia
Phylum	:	Arthropoda
Class	:	Insecta
Order	:	Coleoptera
Superfamily	:	Cucujoidea
Family	:	Coccinellidae
Subfamily	:	Chilocorinae
Genus	:	*Henosepilachna*
Species	:	*vigintioctopunctata*

Symptoms of damage

- The adults are like typical ladybird beetles with wing cases of dull orange and black spots. However, close inspection shows that the upper surface is covered in short downy hairs. The feeding symptom shows scrapping of chlorophyll and the leaf looks like a skeletonization, drying of leaves.

- Adults and larvae strip the surface layers from both sides of the leaves. The damage causes loss of water, and the leaves dry, curl and die.

Biology

Eggs

The eggs are oval, cigar shaped and yellow in colour (1 mm by 0.4 mm) are laid upright in batches of 10-20 on the underside of a leaf. They hatch in about 4 days.

Grub

The pale yellow-whitish larvae have long, dark-tipped branched longitudinal spines on their backs; they grow to 6 mm through three moults in the next 18 days.

Pupa

Yellowish with spines on posterior part and anterior portion being devoid of spines and to attaching themselves to the undersides of the leaves and developing into pupae. This stage lasts another 4 days.

Adult

The adults fall to the ground when disturbed, pretending to be dead. They also produce a yellow fluid that wards off predators. Spread occurs when the adults take to the wing.

Seasonal Incidence

- It was found that minimum temperature morning and evening relative humidity played key role in seasonal population changes of *H. vigintioctopunctata,* among various weather parameters studied.

- Vishav *et al.,* (2017) reported that the grub population reached maximum during 3rd week of August (3.78 grubs/ plant) when the corresponding average maximum and minimum atmospheric temperatures, morning and evening relative humidity and rainfall were 33.37 and 23.74°C, 87.36 and 61.50 % and 4.20 mm respectively.

9. White Grub (*Holotrichia* sp.)

Taxonomy

Kingdom	:	Animalia
Phylum	:	Arthropoda
Class	:	Insecta
Order	:	Coleoptera
Superfamily	:	Scarabaeoidea
Family	:	Scarabaeidae
Subfamily	:	Melolonthinae
Tribe	:	Rhizotrogini
Genus	:	*Holotrichia*
Species	:	*Holotrichia* sp.

Symptoms of damage

- The damage is done mainly by the grubs which feed on roots and tubers in the soil.

- They damage the plant by feeding on the underground portion viz. roots, stems and tubers.

- The early stage feed on the roots with the result the plants dry up and later on when tubers are developed, the grubs cut holes in the tubers.

Biology

Eggs

Eggs are deposited singly in weedy fields or grasslands several centimetres below the soil surface. The time taken for eggs to hatch varies considerably from about 6 days to over 50 days.

Larvae

Larvae undergo three stages (instars). Larvae may also undergo diapause (a period of dormancy triggered by climatic conditions such as dry seasons) prior to pupation. Wet soil conditions during diapause will cause high larval mortality by promoting fungal and bacterial infections.

Pupal

Pupal development takes 30-40 days. Most species of white grub complete their lifecycle in one year; however, one of the most damaging species Phyllophaga implicita can take up to three years to complete its lifecycle.

Adults

Beetles emerge from pupae in the soil in response to the start of the rainy season or soil disturbance (e.g. ploughing). Adults typically emerge at dusk and are active night fliers. Adults commonly feed in trees, but return to grasslands and cultivated fields to lay eggs following mating. Female adults can continue to lay eggs for over 100 days. Up to 60 eggs can be laid at a time.

10. Tobacco caterpillar: (*Spodoptera litura*)

Taxonomy

Kingdom	:	Animalia
Phylum	:	Arthropoda
Class	:	Insecta
Order	:	Lepidoptera
Superfamily	:	Noctuoidea
Family	:	Noctuidae
Genus	:	Spodoptera
Species	:	litura

Damage symptoms

- In early stages, the caterpillars are gregarious and scrape the chlorophyll content of leaf lamina giving it a papery white appearance.
- Later they become voracious feeders making irregular holes on the leaves.
- Irregular holes on leaves initially and later skeletonization leaving only veins and petioles
- Heavy defoliation.

Biology

It is found throughout the tropical and sub-tropical parts of the world, wide spread in India. Besides tobacco, it feeds on cotton, castor, groundnut, tomato, cabbage and various other cruciferous crops.

Egg

Female lays about 300 eggs in clusters. The eggs are covered over by brown hairs and they hatch in about 3-5 days.

Larva

Caterpillar measures 35-40 mm in length, when full grown. It is velvety, black with yellowish-green dorsal stripes and lateral white bands with incomplete ring-like dark band on anterior and posterior end of the body. It passes through 6 instars. Larval stage lasts 15-30 days

Pupa

Pupation takes place inside the soil. Pupal stage lasts 7-15 days.

Adult

Moth is medium sized and stout bodied with forewings pale grey to dark brown in colour having wavy white crisscross markings. Hind wings are whitish with brown patches along the margin of wing. Pest breeds throughout the year. Moths are active at night. Adults live for 7-10 days. Total life cycle takes 32-60 days. There are eight generations in a year.

Seasonal Incidence

- Maximum *S. litura* built up at temperature ranges from 26 to 35.1°C, relative humidity ranges from 89 and 62%, zero rainfall, total sunshine hours (64.6 hrs/week).

- *S. litura* population showes a positive correlation with relative humidity, sunshine hours, whereas negatively correlated with wind velocity.

Biointensive Integrated Pest Management (BIPM)

Biointensive IPM incorporates ecological and economic factors into agricultural system design and decision-making and addresses public concerns about environmental quality and food safety. The benefits of implementing biointensive IPM can include reduced chemical input costs, reduced on-farm and off-farm environmental impacts and more effective and sustainable pest management. An ecology-based IPM has the potential of decreasing inputs of fuel, machinery and synthetic chemicals, all of which are energy intensive and increasingly costly in terms of financial and environmental impact. Such reductions will benefit the grower and society. Over-reliance on the use of synthetic pesticides in crop protection programmes around the world has resulted in disturbances to the environment, pest resurgence, pest resistance to pesticides and lethal and sub lethal effects on non-target organisms, including humans. These side effects have raised public concern about the routine use and safety of pesticides. At the same time, population increases are placing ever-greater demands upon the 'ecological services', that is, provision of clean air, water and wildlife habitat – of a landscape dominated by farms. Although some pending legislation has recognised the costs to farmers of providing these ecological services, it is clear that farmers will be required to manage their land with greater attention to direct and indirect off farm impacts of various farming practices on water, soil, and wildlife resources. With this likely future in mind, reducing dependence on chemical pesticides in favour of ecosystem manipulations is a good strategy for farmers. Biointensive IPM is de fined as 'A systems approach to pest management based on an understanding of pest ecology. It begins with steps to accurately diagnose the nature and source of pest problems, and then relies on a range of preventive tactics and biological controls to keep pest populations within acceptable limits. Reduced-risk pesticides are used if other tactics have not been adequately effective, as a last resort, and with care to minimize risks' (Benbrook 1996). The primary goal of biointensive IPM is to provide guidelines and options for the effective management of pests and beneficial organisms in an ecological context. The flexibility and environmental compatibility of a biointensive IPM strategy make it useful in all types of cropping systems. Biointensive IPM would likely decrease chemical use and costs even further.

1. Potato Tuber Moth

Cultural and Mechanical Management

- Planting seed tubers at a depth of 10 cm as against the traditional planting depth of 6 cm reduce its damage to a great extent (Akhade *et al.,* 1970).

- The fields should be ridged after 6 to 7 weeks of planting so that the tubers are burried at least 25cm below the soil surface.

- In areas where PTM population remains quite high and severe tuber damage is expected, ridging should be done twice so that the tubers are not exposed at any time for egg laying and infestation. Timely and adequate irrigations minimize soil cracking and thereby reduce the risk of tuber exposure to PTM attack or their laying eggs. This problem is quite common in areas where potato crop is taken in heavy soils.

- Harvested tubers must be removed from the field as early as possible and should not be kept overnight in the field.

- The crops like tomato, tobacco, chillies and brinjal should not be grown in the vicinity of potato fields, particularly in PTM prone areas (Lal, 1993).

- Varietal resistance remains the first line of defence in the control of insect pests. Although immunity to PTM has not been identified in potato cultivars so far, yet various levels of genetic resistance/tolerance have been reported in several potato cultivars. An Indian potato hybrid, QB/A 21-29 was found to be tolerant to PTM. Store healthy (PTM free) potatoes in cold stores.

- Covering of healthy tubers stored in country stores with 2-3 cm thick layers of chopped dried leaves of either of Lantana, Soapnut, Neem, Eucalyptus or Eupatorium.

Biological Management

Natural enemies

- In the state of Karnataka, Indigenous parasitoids like *Chelonus curvimaculatus* Cameron, *Bracon gelechiae* Asheamd, *Apanteles sp., Pristomerus vulnerator* Panzer and several other braconids have been found in abundance and are reported to cause 4-17% parasitisation under field conditions (Nair & Rao, 1972).

- A number of exotic parasitoids were also introduced into India between 1970 and 1980. Of these, *Copidosoma koehleri* Blanchard, an exotic egg/larval parasitoid, gave 28.4-60.8% parasitization in Maharashtra (Dalaya and Patil, 1973).

- On the other hand, field releases of *Bracon hebator* Say registered 12% parasitisation of the larvae under field conditions in Bangalore (Divakar and Pawer, 1979).

- Besides, *Orgilus jennieae* Muesebeck, *Apanteles subandinus* Blanch, *Chelonus blackburni* Cameron and *C. kellieae* Cameron, the ovo-larval parasitoids were also tested for their effectiveness in suppressing PTM population at Rajgurunagar (Maharashtra).

Botanical

- Natural insect repellents like leaves, stems, flowers and fruits of various plants possessing repellency properties. Three plant species, viz. *Lantana camara* Linn., *Eucalyptus globulus* L. Herit. and *Minthostachys sp.* either dried/shredded leaves or powder form are effective in preventing PTM damage to potatoes stored for four months in country stores.

- In India, out of the several plant species tested against PTM, *Lantana aculeata* L. provided the best protection by reducing PTM damage from 70% to below the 5%. For this purpose, fresh leaves of lantana are chopped, dried and spread in 2 cm thick layers both below and above the stored potato.

- Dried powders of five different plants, rhizomes of *Acorus calamus,* leaves of *Melia azedarach,* ripen berries of *Piper longum,* leaves of *Prunus persica* and ripen fruit of *Lindera neesiana* were tested for the control of potato tuber moth at laboratory.

- Four natural plant oils (Margorum, Cardamon, Rosemary and Terpintin) were tested against different stages of PTM. The 0.02 and 0.05% concentrations of cardamon oils exhibited the best reduction in percentage of eggs hatchability (67.47 and 86.74%).

- Dusting potato tuber by 1.5% concentrations of cardamom and rosemary oils elicited the lowest percentage of larval penetration, pupation and adult emergence (Moawad and Ebadah, 2007).

Sex pheromones

- PTM sex pheromones consisting of trans-4, cis-7-ol-tridicadien-1 acetate (PTM-1) and trans-4, cis-7-cis-10-tridectrien-1 acetate (PTM-2) in ratio 0.4 mg PTM 1 + 0.6 mg PTM-2 are commercially available from International Potato Center Lima, Peru, Laboratory for Research on Insecticides, Marijkeweg 22,6700 PG, Wageningen, Netherlands and Bhabha Atomic Research Centre, Trombay, Mumbai.

- These sex pheromones are commonly used for monitoring of field population, detecting the occurrence, facilitating proper use of insecticides and mass trapping of PTM for direct control (Chandla *et al.*, 1986).

- These sex pheromones can be used effectively in funnel, water or sticky traps. For monitoring purposes, two traps per field should be installed in the crop 50 metre apart.

- The traps should be checked at 7-day intervals. The number of traps to be used for direct control depends upon the population density of moths, for example, in Lima, population above 2000 moths/trap/day requires 42 traps/ha (Raman & Booth, 1983).

Bacteria

Bacillus thuringiensis Berliner, has also been reported to be effective in controlling PTM in Peru, Tunisia, and India (Amonlar *et al.,* 1979; Chandla *et al.,* 1993; Ranjekaret *et al.,* 2003).

Virus

- A baculovirus has also been collected in Peru and reported as effective against PTM. Such larvae do not move violently when disturbed and also fail to pupate. This virus has been reported quite effective in suppressing the PTM populations both in fields and stores. This virus remains effective up to a period of 120 days under storage condition (Chandla *et al.,* 1993; Lal, 1993).

- Treat potatoes with the dust of granulosis virus before storing the potatoes. For this purpose a mixture in the ratio of 1.0 kg of talc powder + 20 GV infected larvae + 1.0 litre of water + 2 ml of triton should be prepared and used @ 5 kg of mixture/ ton of potatoes. 184 Potential Plant Protection Strategies.

CIPC

- CIPC (Isopropyl N-(3-chlorophenyl) carbamate) is a sprout suppressant commonly used on ware potatoes in Country stores. But beside suppression of sprouts.

- CIPC was found to reduce PTM infestation. After the period of 60 days of incubation only 2-6% tuber damage was observed in the country stores (Chandla *et al.,* 2003).

2. Cutworms

Cultural and Mechanical Management

- Cutworms either aestivate during summer months or hibernate during winters in the soil while completing their life cycles. Therefore, deep ploughing of potato fields during summer months

- In the plains exposes the immature stages to high temperature and predatory birds (Chandla, 1985; Mishra & Agarwal, 1988; Mishra *et al.,* 1995). Similarly, deep ploughing of fields during autumn in the hills also minimises cutworm's population.

- Light traps installed in/around potato fields attract the adults of cutworms, and helps in mass collection and destruction of the moths.

Botanical

- Garlic as intercrop with potato was found to be effective in minimizing cutworm damage in potato crop at Shimla.

Natural enemies

- Broscus punctatus Dist and Liogryllus bimaculatus Linn have been reported to be the parasitizes of cutworm larvae (Fletcher, 1916). Macrocentrus collaris Spin, Netelia ocellaris Thomson, Coelichneumon sp. nr truncatulus Thomson, Periscepsia carbonaria Panzer and Turanogonia chinensis Wiedemann also parasitise A. ipsilon and A. segetum under natural field conditions (Chandla et al., 1989; Singh, 1993).

Bacteria

- Bacillus thuringiensis Berliner is a well-known biopesticide. Spraying the crop and ridges with this biopesticide (Bt @ 109 spores/ml) gives a good control (Mishra et al., 1995).

Fungus

- Entomogenous fungus, *Metarrhizium anisopliae* Meld. is a best known fungal control of cutworm (Mishra et al., 1995).

Nematodes

- Entomophilic nematode, Stinernema (Neoaplectana) sp. are also well known as dominant regulatory factors for cutworm populations from various parts of the country (Mishra et al., 1995).

3. Aphids

Cultural and Mechanical Management

- Growing potato crop by adopting the seed plot technique with the following precautions, i.e., planting of clean (virus free) potato seed purchased from certified/reliable sources; growing of seed crop during aphid free/low aphid periods; maintaining an isolation of 50 m for seed crop; timely rouging of virus infected plants well before they touch each other; haulm (foliage) cutting of crop after desired maturity of crop but before the aphids cross the critical level of 20 aphids/100 compound leaves and (Verma *et al.*, 1993a); Regular cutting of foliage regrowth so that the aphids could not build up on them.

- Application of mineral oil (Ferro *et al.* 1980, Lowery *et al.* 1990) and use of aluminum or white plastic mulch (Wyman *et al.* 1979) reduce virus transmission.

- Aphids that are not effectively repelled by reflective mulch seem to thrive on mulched crops (Zalom, 1981) and exhibit high rates of reproduction. Therefore, even in mulched crops some aphid control is necessary.

Biological Control

Natural Enemies

Over 24 predators and 22 parasitoids reportedly attack Aphid, M. persicae (Singh, 1988). However, very few of these are effective:

- At Shimla, *Aphelinus* sp. has been found to parasitize 100 per cent *M. persicae* under glasshouses conditions. Similarly, on *A. gossypii,* over 36 predators and 21 parasitoids have been recorded (Singh, 1988).

- In Karnataka, Acolemani parasitized upto 70% aphids in certain potato fields (Trivedi, 1988). At Shimla, one Aphelinus has showen its maximum capability by parasitizing 100% of *M. persicae* alone (Verma *et al.,* 1976a).

- Seven species of coccinellids, two of syrphids and a chryopid were recorded on potato around Farrukhabad, Agra and Meerut districts of Uttar Pradesh. Among these, *Cooccinella septempuncata* Linn. and *M. sexmaculatus* Fabr. Were found predominant (Mishra *et al.,* 1995).

- Ten species of cocccinella, two chrysopids and three predaceous bugs were recorded feeding on aphids in Shimla during May-October (Anon, 1984). Five entomopathogenic fungi were reported against aphids in Shimla (Singh, 1988).

4. Whitefly

Cultural control

- Remove weeds and crop residues to avoid infestation
- Mulch the soil with straw
- Use yellow traps to attract adults
- Spraying soap solution with sticker-spreader can reduce whitefly population wasp

- Conserve natural enemies (green lacewing, minute pirate bug, ladybug, parasitic wasp)
- Yellow sticky traps are helpful for monitoring and suppressing adult populations.
- If found, use the Bug Blaster to hose off plants with a strong stream of water and reduce pest numbers.

Mechanical controls

- Traps offer a pesticide free method of control of *B. tabaci*. The Light-Emitting Diode Equipped CC trap (LED-CC) was developed by plant physiologist Chang-Chi Chu and Thomas Henneberry. The trap itself includes a green LED light that attracts and traps the whiteflies. The LED device works best at night, and is inexpensive and durable. In addition, the LED is does not harm predators and parasitoids of the whitefly.

Trap crops

- Squash crops are effectively used as trap crops for attracting silverleaf whitefly. Another important control is the use of other crops as a source of trap crops. Squashes can act as trap crops for the silverleaf whitefly due to the flies' attraction to these crops.

Biological control

- Natural predators of this pest include ladybugs and lacewing larvae, which feed on their eggs and the whitefly parasite which destroys nymphs and pupae. For best results, make releases when pest levels are low to medium
- Safer® Soap will work fast on heavy infestations. A short-lived natural pesticide, it works by damaging the outer layer of soft-bodied insect pests, causing dehydration and death within hours. Apply 2.5 oz/ gallon of water when insects are present, repeat every 7-10 day as needed.

Botanical control

- BotaniGard ES is a highly effective biological insecticide containing Beauveria bassiana, an entomopathogenic fungus that attacks a long-list of troublesome crop pests – even resistant strains! Weekly applications can prevent insect population explosions and provide protection equal to or better than conventional chemical pesticides.
- Organic Neem Oil can be sprayed on vegetables, fruit trees and flowers to kill eggs, larvae and adults. Mix 1 oz/ gallon of water and spray all leaf surfaces (including the undersides of leaves) until completely wet.

- Horticultural oils, which work by smothering insects, are very effective on all stages of this pest.

- Fast-acting botanical insecticides should be used as a last resort. Derived from plants which have insecticidal properties, these natural pesticides have fewer harmful side effects than synthetic chemicals and break down more quickly in the environment.

5. Thrips: (*Thrips tabaci*)

Cultural control

- Sanitation is the first and most important step in implementing an effective pest management program.

- Effective sanitation will reduce or even eliminate thrips as a pest problem. For example, in cut roses, removing all flower buds (including non-marketable flowers) can significantly reduce thrips populations in that crop.

- Cultural control measures also include maintaining a healthy crop and an optimal greenhouse environment (such as 80% relative humidity), creating less favourable conditions for a rapid increase in the density of thrips populations.

Biological control

Biological control agents include predatory mites such as

- *N. cucumeris* and *A. swirskii* are the most extensively used predatory mites and look very similar. These mites control western flower thrips on the foliage by feeding on the first instar larvae. A. swirskii can also feed to a lesser extent on second instar thrips.

- Orius is effective in controlling thrips. Unlike *N. cucumeris* and *A. swirskii*, *Orius* will feed on all stages of thrips.

- Stratiolaelaps scimitus and Gaeolaelaps gillespiei are soil-dwelling predatory mites that feed on a variety of soil organisms, including thrips pupae.

Botanicals

- *Beauveria bassiana* and *Isaria fumosorosea*). is a fungal pathogen of thrips. It is usually mixed in water and applied as a spray. In vegetables, it can be either sprayed onto the crop or distributed via bumble bees that are supplied with hives specially equipped with dispensing trays.

6. Leaf minor: (*Liriomyza* spp.)
Biological Control

- The parasitoids of *Liriomyza sativae* (as well as *Liriomyza trifolii* and *Liriomyza huidobrensis*) often display little specificity. The wasp parasitoids often attack all three species, and when they appear to be specific, it is usually lack of knowledge about host range rather than actual specificity (Murphy and LaSalle, 1999).

- The most common predators are mirids, including: *Cyrtopeltis modestus* (Dist.), *Dicyphus cerastii* Wagner, *Dicyphus tamaninii* Wagner and *Macrolophus caliginosus* Wagner. The adults and nymphs are mobile and can prey on leafminer larvae or pupae. D. tamaninii may also damage tomato fruits when prey density is low.

- *M. caliginosus* originates from the Mediterranean, is an important predator of *Liriomyza*, and is able to survive even with low levels of food. *M. caliginosus* used alone or in combination with parasitoids (*D. isaea*) has been used to control *L. bryoniae* in commercial tomato greenhouse situations.

- A ponerine ant (Formicidae: Ponerinae) has been recorded attacking *L. trifolii* larvae. A cecidomyiid fly, *Aphidoletes aphidimyza* (Rondani) (Diptera: Cecidomyiidae) has been recorded as a predator of *L. bryoniae* on tomato in greenhouses.

- A lynx spider in the Oxyopidae (Arachnida) has been recorded attacking *L. trifolii* adults.

Cultural practices

- Some crops vary in susceptibility to leaf mining. This has been noted, for example, in cultivars of tomato, cucumber, cantaloupe, and beans (Hanna *et al.*, 1987).

- Nitrogen level and reflective mulches are sometimes said to influence leafminer populations, but responses have not been consistent (Chalfant *et al.*, 1977, Hanna *et al.*, 1987).

- Placement of row covers over cantaloupe has been reported to prevent damage by leafminer (Orozco-Santos *et al.*, 1995).

Entomopathogens

Borisov and Ushchekov (1997) tested six species of entomopathogenic fungi against *L. bryoniae,* and found that *P. lilacinus* and *M. anisopliae* were the most effective, reducing adult emergence from soil.

Entomopathogenic Nematodes

- Harris MA, Begley JW, Warkentin (1990) reported that few species of entomopathogenic nematodes have been found infecting *Liriomyza spp.*, and those nematodes include *Heterohabditis heliothidis, Heterohabditis megidis, Heterohabditis sp.* (strain UK 211), *Steinernema carpocapsae* (Weiser) and *Steinernema feltiae* (Filipjev) (=*Neoaplectana feltiae*)

7. Jassid: *Amrasca biguttula* (Ishida)

Natural enemies of leaf hopper

Parasitoids: *Anagrus flaveolus, Stethynium triclavatum* etc.

Predators: Lacewing, red ant, mirid bug, big-eyed bug, ladybird beetle etc.

8. Epilachna Beetles

Cultural control

- Do not plant potato next to crops that are known to be alternative hosts of the hadda beetle, other members of the Solanaceae, including weeds, and also beans.

- Handpick the larvae, and perhaps the adults. If attempted, it should be done when the beetles are first seen in the crop.

- Remove weeds in the Solanaceae family from around the crop. However, it has been suggested these might.

- Collect crop debris after harvest and burn it.

Biological Management

Natural enemies

- A eulophid egg parasitoid, *Tetrastichus ovulorum* Ferriere, was recorded on *E. vigintioctopunctata* from Bangalore and Ranchi (Krishnamurti, 1932; Lall, 1964). A chalcid parasitoid, Ugna menoni Kerrich was recorded on *H. vigintioctopuncta* from Andhra Pradesh (Azam *et al.,* 1974) and *H. ocellata* from Shimla hills (Saxena and Singh, 1982).

- The parasitisation of H. vigintioctopunctata in the field by Pediobius foveolatus Crawford has been reported to be as high as 77 per cent.

- Two reduviids, Coranus sp. and Henricoclania sp. nr. spinosa Dist. and a nabid, Aptus mussooriensis Dist. have been found feeding on the grubs and adults of H. ocellata at Shimla (Saxena and Singh, 1977) and Cantheconidea fusicellata Wolf. from Kerala on H. vigintioctopunctata (Shiella and Abraham, 1981).

Botanicals

Effects of leaf extracts on life-history traits

- The oviposition of *E. vigintioctopunctata* females reared on the host leaves treated with the plant extracts reduced significantly compared to the control females. The highest reduction in egg-laying was recorded for *Calotropis procera* in a dose-dependent manner.

- The hatchability of eggs laid on treated host leaves reduced significantly from about 68%, indicating a dose-dependent inhibition in egg-hatching in *E. vigintioctopunctata*. Similar to oviposition deterrent, the most potent extract inhibiting egg-hatch was of *Ricinus communis*. Relatively poor inhibition was recorded for *Datura metel* and *C. procera*.

- The overall larval duration was significantly prolonged in the treated lines, where R. communis also had the maximum effect. However, differences among the control and other two treatments were not statistically significant.

- The formation of pupae in *E. vigintioctopunctata* was significantly inhibited by the leaf extracts of *R. communis* and *C. procera* showed much more drastic effects than that of *D. metel.*

- *R. communis* and *C. procera* had more pronounced inhibition effects on adult emergence in *E. vigintioctopunctata.*

- At nine days after spraying, among the biopesticides, azadirachtin, the botanical insecticide was found very effective against the epilachna beetle followed by the Pongamia at the higher concentration.

Microbial control

- The microbial pesticide *B. bassiana* provided only 39.56 % suppression of the epilachna beetle population.

9. White Grub

Cultural and Mechanical Management

- In the endemic areas, autumn ploughing in hilly areas not only exposes the grubs to adverse conditions (low temperature) but the exposed grubs also become prey of the birds.

- The use of nitrogenous fertilizers, especially ammonia and urea, at higher doses kill the first instar grubs.

- The light traps may also be used for collecting the beetles during night. The beetles can also be collected by shaking or jerking the host plants during night.

- The fallen beetles should be collected and destroyed by putting them either in kerosinized water or by burning.

- The host trees of adults (beetles) should be lopped or pruned.

Water Management

- Practically all white grub species require moist soil for their eggs to hatch. The young larvae are also very susceptible to desiccation.

- In areas which stand some moisture stress, do not water in July and early-August when white grub eggs and young larvae are present.

- On the other hand, moderate grub infestations can be outgrown if adequate water and fertilizer is applied in August through September and again in May when the grubs are feeding.

Host Plant Modifications

- Certain species of scarab adults prefer specific host plants. Where beetles are common, do not plant roses, grapes and lindens around high maintenance turf areas.

- May/June beetles prefer oaks and the green June beetles feed on ripening fruit such as peaches. The fine and tall fescues are not as severely attacked as perennial rye grass.

Biological Management

- A number of bio-control agents are known to manage white grubs in different parts of the country. These should be encouraged either by conserving the existing populations or by introducing and establishing the known bio-control agents obtained from new localities.

Botanical

- In India, out of the several plant species tested against grubs, Neem seed kernal and Eucalyptus globulus in powder form or in solution made in water are effective.

Natural enemies

- Some of the potential parasites scoliids, Campsomeris collaris F., Scolia aureipennis Lepeletier and S. pustulata Magr. are known to be effective against white grubs.

Bacteria

• Bacillus sp is a well-known biopesticide. Spraying the crop and ridges with this biopesticide Bacillus thuringiensis (Bt @ 109 spores/ml) gives a good control (Mishra *et al.*, 1995).

• Several strains of the bacterium, Bacillus popilliae, have been found that attack white grubs. However, the commercial preparation of this bacterium is extracted from Japanese beetle grubs and are most active against this species.

• This bacterium is picked up by feeding grubs and it causes the body fluids to turn a milky-white before grub death. Fresh bacterial preparations should be used and three to five years are needed to provide lasting controls.

Fungus

• Entomogenous fungus, Metarrhizium anisopliae Meld is best known fungal control of white grubs (Mishra *et al.*, 1995). Beauveria brongniartii Sacc. is also very effective against white grubs.

10. Tobacco caterpillar: (*Spodoptera litura*)

Biological control

Egg Parasitoids

• Four species of trichogrammatids, one scelionid and one braconid which had been reported as egg parasitoids of S. litura, an unidentified Chelonus species and species of Telenomus, have also been reported as both egg and larval parasitoids.

• A total of 10 egg parasitoids have been reported from different parts of the host distribution. Among the trichogrammatids, T. chilonis from India (Joshi *et al.*, 1979; Patel *et al.*, 1971) and T. dendrolimi from China (Chiu and Chou, 1976) are the most common. These species are often reported from the eggs of several other hosts.

Larval Parasites

• In India, 32 different species of parasitoids have been reported as larval parasitoids of S. litura. Among these, Apanteles and species of Bracon were the most commonly reported.

• Rai (1974) surveyed vegetable crops in the state of Karnataka and found that 10% of larval mortality was caused by Chelonus formosanus.

- Jayanth and Nagarkatti (1984) reported the emergence of up to 12 tachninid parasitoids (Peribaea orbata) from a single S. litura larva in Karnataka state, India.

- Rao and Satyanarayana (1984), during a pest survey of natural enemies of S. litura in Andhra Pradesh, India, reported *Zele chlorophthalma* as a larval parasitoid.

Pupal Parasitoids

- Relatively few pupal parasitoids have been reported from *S. litura*. Eight parasitoid species have been reported from the pupal stage of S. litura, one of which is a larval-pupal parasitoid (Ichneumon sp.) and one a prepupal parasitoid (species of Chelonus).

Predators

- Altogether 36 predatory insects from 14 families and 12 species of spiders, representing six families were reported to feed on S. litura eggs, larvae and pupae in different parts of the world. Of the total predators reported to feed on S. litura, 50% of the insect predatory fauna and 83% of the spiders were from India.

- The biology of Canthoconidia furcellata was studied in the laboratory with a view to using this predator in an integrated pest management programme for tobacco pests. Chu and Chu (1975) studied the effects of temperature on the growth of C. furcellata and found that 71,216 and 134 degree days were required for egg, nymph and adult stages, respectively. It was concluded that there are five to six generations per year of this predator in northern Taiwan.

- Nakasuji *et al.* (1976) observed a predatory wasp, preferentially selecting fifth- and sixth-instar larvae over early instars. The wasps were more active and attacked more larvae in fields with high larval density than those with low larval density. However, the percentage of predation was lower in the field with highest density of S. litura larvae.

- Deng and Jim (1985) reported Conocephalus sp. as a new predator on egg masses of S. litura in Guanxi, China. This katydid was successfully reared on an artificial diet. Field releases of nymphs and adults of Conocephalus sp. were attempted for control of Scirpophaga incertulus.

11. Broad Mite

Cultural control

* Suitable crop rotations with non-host crops like wheat in pest prone areas and providing proper isolation to potato crop from susceptible hosts like chillies and brinjal.

* Delayed planting undertaken in last week of September for early crop and during middle of October for the main crop reduces the incidence and impact of the pest.

References

Adams S (1997). Seein' red: colored mulch starves nematodes. Agricultural Research. October, p 18

Akhade, M.N., Tidke, P.M. and Patkar, M.S. (1970). Control of potato tubemoth (*Gnorimoschema operculella* Zell.) in Deccan Pleatue through insecticides and depth of planting. Indian *J. Agri. Sci.* 40: 1071-1076.

Amonkar, S.V., Pal, A.K., Vijaylakshmi, L. and Rao, A.S. (1979). Microbial control of potato tuber moth (Phthorimaea operculella Zell.). *Indian J. Expetl. Biol.* 17: 1127-1133.

Anandraj M, Eapen SJ (2003). Achievements in biological control of diseases of spice crops with antagonistic organisms at Indian Institute of Spices Research, Calicut. In: Ramanujam B, Rabindra RJ (eds) Current status of biological control of plant diseases using antagonistic organisms in India. Project Directorate of Biological Control, Bangalore, pp 189–215

Anonymous (1984). Quinnquennial review team reports, AICRP on biological control of pests and weeds. *Tech. Doc. No.* 6: 141.

Azam, K.M., Aziz, S.A. & Ali, M.M. (1974). A new record of Ugna menoi Kerrich as parasite of *Epilechna vigintioctopunctata* (F) in *India. Curr. Res.* 3: 88.

Benbrook CM (1996). Pest management at the crossroads. Consumers Union, Yonkers, 272 pp

Bombawala, O.M., Bedi, P.S., Sharma, N.K. and Sharma, S.K. (1982). B.B. Nagaich In: Developing Countries. Indian Potato Association CPRI, Shimla. pp. 425-427.

Borisov B.A., Ushchekov A.T. (1997). Entomogenous fungi – Hyphomycetes against the nightshade leaf miner. *Zashchita i Karantin Rastenii*; 5: 10–11.

Chalfant R.B., Jaworski C.A., Johnson A.W., Summer D.R. (1977). Reflective film mulches, millet barriers, and pesticides: effects on watermelon mosaic virus, insects, nematodes, soil-borne fungi, and yield of yellow summer squash. *Journal of the American Society of Horticultural Science* 102: 11-15.

Chandla V.K., Singh, B. and Chandel, R.S. (2003). Management of Potato tuber moth in countary stores with CIPC. *J. Indian potato Assoc* 30(1-2): 153-154.

Chandla, V.K. (1983). Extent of damage by pests. B.B. Nagaich (Eds.). In: potato production storage and Utilization. Central Potato Research Institute, Shimla. pp. 425-427.

Chandla, V.K. (1985). Potato pest and their management. *Indian farming* 36(12): 31-32.

Chandla, V.K. (1986). Insect pest complex of potato crop in Shimla hills and their management. Ph.D Thesis. Department of entomology–Apiculture, Dr. Y.S. Parmar University of Horticulture and Forestry, Solan (H.P.) pp. 94.

Chandla, V.K., Mishra, S.S. and Verma, K.D. (1989). Natural enemies of Agrotis species in Shimla hills (Abstract). Presented in "Xth Annual., General Body Meeting of IPA., held at HAU, Hisar on 20th Aug. 1989.

Chandla, V.K., Mishra, S.S., Sharma, D.C. and Kashyap, N.P. (1993). Potato tuber moth menece in Himachal Pradesh and its management (Abstract). *J. Indian potato Assoc* 20(1): 60-61.

Chaudhary, R., Trivedi, T.P. and Raj, B.T. (1983). Field evaluation of some exotic parasitoid of Potato tuber moth (*Phthorimaea operculella Zell.*). *Indian J. Ent.* 45: 504-506.

Couch GJ (1994). The use of growing degree days and plant phenology in scheduling pest management activities. Yankee Nursery, Quarterly Fall, pp 12–17.

Dalaya, V.P. and Patil, S.P. (1973). Laboratory rearing and field releases of Copidosoma koehleri Blachard, an exotic parasite for the control of Gnorimoschema operculella Zeller. *Res. J. Mahtma Phule Agri. Univ* 4: 97-107.

Diwakar, B.J. and Pawar, A.D. (1979). Field recovery of Chelonus blackburni and Bracon hebtor from Potato tuber moth. *Indian J. Plant Pro.* 7: 214.

Fletcher, T.B. (1916). Annotated list of Indian crop pests. *Rep Agri. Res. Inst and Col. Pusa* 1915. pp. 58-77.

Ghosh, D., Nath, D. and Chakrabarti, S. (1985). Predators and parasites of aphids (Hom., *Aphididae*) from North-West Himalaya: Ten species of syrphids (Diptera-*Syrphidae*) from Garhwal range. *Entomon* 10(4): 301-303.

Hanna H.Y., Story R.N., Adams A.J. (1987). Influence of cultivar, nitrogen, and frequency of insecticide application on vegetable leafminer (Diptera: Agromyzidae) population density and dispersion on snap beans. *Journal of Economic Entomology* 80: 107-110.

Harris M.A., Begley J.W., Warkentin D.L. (1990). Liriomyza trifolii (Diptera: Agromyzidae) suppression with foliar applications of Steinernema carpocapsae (Rhabditida: Steinernematidae) and abamectin. *Journal of Economic Entomology*; 83: 2380–4.

Kamrul Islam1, M.S. Islam and Zennat Ferdousi (2011). Control of *Epilachna vigintioctopuntata* FAB. (Coleoptera: *Coccinellidae*) using Some Indigenous Plant Extracts. *J. Life Earth Sci.,* Vol. 6: 75-80.

Kapoor, A.P. (1950). A note on Epilachna ocellata Redt. (Coleoptera: *Coccinellidae*) with description of three species hitherto confused with it. *Rec. Indian Mus.* 48: 17-19.

Khaderkhan H, Nataraju MS, Nagaraja GN (1998). Economics of IPM in tomato. In: Reddy PP, Kumar NKK, Verghese A (eds) Advances in IPM for horticultural crops. Association for Advancement of Pest Management in Horticultural Ecosystems, Division of Entomology and Nematology, Indian Institute of Horticultural Research, Bangalore, pp 151–152.

Krishna Moorthy P.N., Krishna Kumar N.K. (2002). Advances in the use of botanicals for the IPM of major vegetable pests. In: Proceedings of the international conference on vegetables , Bangalore. Dr. Prem Nath Agricultural Science Foundation, Bangalore, pp 262–272.

Krishna Moorthy P.N., Krishna Kumar N.K., Girija G., Varalakshmi B., Prabhakar M. (2003). Integrated pest management in cabbage cultivation. Extension Bulletin No. 1, Indian Institute of Horticultural Research, Bangalore, 10 pp

Krishnamurti, B. (1932). The potato epilachna beetle, *Epilachna vigintioctopunctata* (Fabr.). Bull. Dept. Agric. *Mysore Ent. Ser.* 9: 16.

Lal, L. (1993). Potato tuber moth-Bionomics and Management. In: Advances in Horticulture-potato, Eds. Chandha, K.L. and Grewal, J.S. Malhotra Publishing House, New Delhi. 7: 591-602.

Lall, B.S. (1964). Vegetable pests. In: Entomology in India. (Eds). N.C. Pant. Entomological Soc. India. pp. 184.

Misra, S.S. and Agarwal, H.O. (1988). Potato pests in India and their control. Trop. *Pest manag* 35: 199-209.

Misra, S.S. and Chandla, V.K. (1989). White grubs infesting potatoes and their management. J. Indian Potato Assoc. 16(1&2): 29-33.

Misra, S.S., Chandla, V.K. and Singh, A.K. (1995). Potato pests and their management. Technical Bulletin No. 45. Central Potato Research Institute, Shimla. pp. 57.

Moawad, S.S. and Ebadah, I.M.A. (2007). EbadahImpact of Some Natural Plant Oils on Some Biological Aspects of the Potato Tuber Moth, Phthorimaea operculella, (Zeller) (Lepidoptera : Gelechiidae). *Res. J. Agri. Biol. Sci.* 3(2): 119-123.

Nagaich, B.B., Verma, K.D. and Upreti, G.C. (1970). Hereditary variation in the ability of Myzus persicae to transmit potato leaf roll and virus Y. Final Tech. Rept. PL 480 Scheme, CPRI.

Nair, R. and Rao, V.P. (1972). Results of the survey for natural enemies of potato tuber moth, Phthorimaea operculella Zeller (Lepidoptera : Gelechiidae) in Mysore state and the parasites reared from Maharashtra state. *Tech. Bull. Commonwealth Inst. Biolcontro.*, 15: 115-130.

Niroula, S.P. & Kamini Vaidya (2004). Efficacy of Some Botanicals against Potato tuber moth, Phthorimaea operculella (Zeller, 1873) *Our Nature* 2: 21-25.

Orozco-Santos M., Perez-Zamora O., Lopez-Arriaga O. (1995). Floating row cover and transparent mulch to reduce insect populations, virus diseases and increase yield in cantaloupe. *Florida Entomologist* 78: 493-501.

Parvatha Reddy P., Nagesh M., Devappa V. (1997). Effect of integration of Pasteuria penetrans, Paecilomyces lilacinus and neem cake for the management of rootknot nematode infecting tomato. *Pest Managmt Hortil Ecosystems* 3: 100-104.

Parvatha Reddy P., Rao M.S., Nagesh M. (2002). Integrated management of burrowing nematode (Radopholus similis) using endomycorrhiza (Glomus mosseae) and oil cakes. In: Singh HP, Chadha KL (eds) Banana. AIPUB, Trichy, pp 344–348.

Prasad, K.S.K. (1993). Nematodes-distribution, biology and management. In: Advances in Horticulture: Potato. Eds. K. L. Chadha and J. S. Grewal. Malhotra Publishing House, New Delhi. 7: 635-647.

Pruthi, H.S. (1946). Report of the Imperial Entomologist. Abrigd. Sci. Rep. Agric Res. Inst. New Delhi. pp. 64-71.

Raman, K.V. and Booth, R.H. (1983). Evaluation of technology for integrated control of potato tuber moth in field and storage. Technology Evaluation Series No. 10. International Potato Center, Lima, Peru. pp. 18.

Ranjekar P.K., Aparna Patankar, Vidya Gupta, Raj Bhatnagar, Jagadish Bentur and Ananda Kumar, P. (2003). Genetic engineering of crop plants for insect resistance *Cur. Sci.* 84: 321-329.

Reichert S.E., Leslie B. (1989). Prey control by an assemblage of generalist predators: spiders in garden test systems. *Ecology Fall*, pp 1441-1450.

Sarma Y.R. (2003). Recent trends in the use of antagonistic organisms for the disease management in spice crops. In: Ramanujam B, Rabindra RJ (eds) Current status of biological control of plant diseases using antagonistic organisms in India. Project Directorate of Biological Control, Bangalore, pp 49-73.

Saxena, A.P. (1981). Leucopis fumidilarva Tanas (*Diptera*: Chamaemycides), a new predator of aphids and leaf hoppers on potato crop. *J. Indian potato. Assoc* 8(2): 96-98.

Saxena, A.P. and Raj, B.T. (1980). Studies on the biological control of potato pests. Proc. III. Workshop on Biol. Control of crop pests and weeds. PAU, Ludhiana, October 27-30, pp. 121-125.

Saxena, A.P. and Singh, V. (1977). Predacious bugs on Epilachna ocellata Redt. (Coccinellidae: *Coleoptera*). *J. Indian Potato Assoc.* 4: 70.

Saxena, A.P. and Singh, V.P. (1982). Natural enemies of potato pests in India. In: Potato in developing countries, Ed.B.B. Nagaich. Central Potato Research Institute, Shimla. pp. 349-355.

Seshadri, A.R. and Sivakumar, C.V. (1962). Golden nematode of potatoes (Heterodera rostochiensis Woll.), a threat to potato cultivation in the Nilgris. Madras Agric. J. 49: 281-288.

Sheila, M.K. and Abraham, G.C. (1981). Cantheconidea furcellata (Wolf.) (Pentatomidae:*Hemiptera*), as a predator of Henosepilachna vigintioctopunctata (Fabr.) *Agric. Res. J. Kerala* 19: 142.

Singh Amerika, Trivedi T.P., Sardana H.R., Sabir N., Krishna Moorthy P.N., Pandey K.K., Sengupta A., Ladu L.N., Singh DK (2004) Integrated pest management in horticultural crops – a wide area approach. In: Chadha KL, Ahluwalia BS, Prasad KV, Singh SK (eds) Crop improvement and production technology of horticultural crops. Horticulture Society of India, New Delhi, pp 621–636.

Singh, M.N., Khurana, S.M.P., Nagaich, B.B. and Agarwal, H.O. (1982). Efficiency of Aphis gossypii and Acyrthosiphon pisum in transmitting potato viruses X and leafroll. In: Potato in Developing countries. (Eds.). Nagaich, B.B., Shekhawat, G.S., Gaur, P.C. and Verma, S.C. Central Potato Research Institute, Shimla. pp. 289-293.

Singh, S.P. (1974). Observations on the biology of Trichoplusia orichalcea Fabr. on potato. *Indian. J. Plant Prot.* 2(1&2): 127-128.

Singh, S.P. (1988). Natural enemies of aphids. Fourth National Symposium of Aphidology, CPRI, Shimla. October 4-6: pp. 66.

Singh, S.P. (1993). Biological control of potato pests. In: Advances in Horticulture: Potato. Vol. 7. Malhotra Publishing house, New Delhi, 11006 (Eds.). Chadha, K.L. and Grewal, J.S. pp. 615-634.

Srinivasan K., Krishna Moorthy P.N. (1991). Indian mustard as a trap crop for management of major lepidopterous pests on cabbage. *Trop Pest Mang* 37: 26–32.

Srinivasan K., Krishna Moorthy P.N., Raviprasad T.N. (1994). African marigold as a trap crop for the management of the fruit borer Helicoverpa armigera on tomato. *Int J Pest Mang* 40: 56–63.

Sunil Kr. Ghosh and Gautam Chakraborty (2012). Integrated field management of Henosepilachna vigintioctopunctata (Fabr.) on potato using botanical and microbial pesticides, Management of *Henosepilachna vigintioctopunctata, JBiopest, 5* (supplementary): 151-154.

Swaminathan M.S. (2000). For an evergreen revolution. The Hindu Survey of Indian Agriculture 2000: 9-15.

Thirumalachar, M.J. (1951). Root-knot nematode in potato tubers in Shimla. *Curr. Sci.* 20: 104.

Trivedi, T.P. (1988). Natural enemies of potato pests in Karnataka. *J. Indian Potato Assoc.* 15(3&4): 159-160.

Trivedi, T.P. (1990). Spatial distribution, phenology, life table and assessment of losses due to potato tuber moth. Phthorimaea operculella (Zeller) Ph.D. Thesis of University of Agriculture Sciences. Banglore (India). pp. 182.

Verma, K.D. (1985). Potato aphids with special reference to Myzus persicae (Sulzer). Proc. 2nd Nat. Symp. On recent trends in Aphidological studies. (Eds). S.P. Kurl pp. 39-46.

Verma, K.D. and Chauhan, R.S. (1993a). The life cycle of potato vector, Myzus persicae (Sulzer). *Curr. Sci.* 65: 488-89.

Verma, K.D., Misra, S.S. and Saxena, A.P. (1976a). A role of Aphelinus sp. in the natural control of Myzus persicae on potato. *J. Indian Potato. Assoc.* 3(1): 40.

Verma. K.D., Parihar, S.B.S. and Khan, I.A. (1993b). Field reaction of some potato varieties/germplasm lines to leafhopper, *Empoasca fabeae* (Harris). *J. Indian Potato. Assoc.* 20: 49.

Vishav Vir Singh Jamwal, Hafeez Ahmad, Ashutosh Sharma, Devinder Sharma (2017). Seasonal abundance of *Henosepilachna vigintioctopunctata* (Fab.) on *Solanum melongena* L. and natural occurrence of its two hymenopteran parasitoids, Braz. arch. biol. technol. vol.60 Curitiba 2017 Epub May 18.

Williams E.C., Walters K.F.A. (2000). Foliar application of the entomopathogenic nematode Steinernema feltiae against leafminers on vegetables. *Biocontrol Science and Technology;* 10: 61-70.

15

Biointensive Integrated Pest Management of Chilli

Dhole, R.R., Patil, S.P., Mahalle, R.M.and Maurya, R.

Introduction

Chilli (*Capsicum annuum* var.*annuum* L.) is one of the most important economical and popular vegetable crops grown for its green fruits as vegetable and red one as a spice[1]. It is a native of southern Texas and Maxico and was introduced in India by the Portuguese in 19th century[2]. Chilli is mainly used in culinary adding flavor, color, pungency and rich source of vitamins like A, C, E, P and having medicinal properties[3]. Chilli is one of the important cash crops grown in almost all parts of the country and is widely grown in the tropics and subtropics as also under glass houses in temperate regions[4]. The plants are very sensitive to excessive rainfall, water lodging and frost[2,5]. The ideal condition for its cultivation are well drained loam soil rich in organic matter but can also be grown in many type of soils[6]. India is a major producer, exporter and consumer of chilli[7]. Chilli is presently grown extensively throughout the country, both under rainfed and irrigated conditions, in almost all the states covering an area of 792.1 thousand Ha with annual production of 1223.4 thousand metric tonnes[8]. Andhra Pradesh is the largest producer of chilli in India. It contributes about 30% to the total area under chilli, followed by Karnataka (20%), Maharashtra (15%), Orissa (9%), Tamil Nadu (8%) and other states contributing 18% to the total area under chilli[8].

The insect pests cause significant damage to the chilli crop[9]. There are 39 genera and 51 species of insects and mite attacking chilli in the field and in the storage[10]. Aphids (*Aphis gossypii* Glover), Thrips (*Scirtothrips dorsalis* Hood) and Jassids (*Amrasca bigutulla bigutulla*) are the major insect pest of chilli[11]. These pests cause serious damage to the chilli crop by direct feeding and transmitting deadly disease called "leaf curl disease" or "Murda complex"[13]. In addition to these, pod borers also cause maximum damage to the crop both during vegetative and fruit formation stages[12]. The crop loss by three major pests, where 30-50% by thrips (*Scirtothrips dorsalis*), 30-70% by mites (*P. latus*) and 30-40% by fruit borers *Helicoverpa armigera* and *Spodoptera litura*[13]. Among thrips, *Scirtothrips dorsalis* Hood (Thripidae: Thysanoptera) is one of the most destructive pest of chilli and under severe infestation 30 to 50 percent crop may be lost[11]. Thrips alone is reported to be a major pest of chilli in south India as well as in some northern parts of country and there is a need to manage these pests effectively and economically in chilli crop[12]. Insect pests are very destructive to the production of chilli crop[13]. Thus to attain marginal or above marginal production of chilli, controlling the pest below EIL (Economic Injury Level) is very essential. For this biointensive integrated control measure will suits more effectively than any other control measures like chemical one due to their other side effects or problem in their implementation. Therefore, here we are going to discuss in brief about pest and their integrated biointensive control measures.

Major pests of chilli

The major insect pests of chilli crop are discussed along with their Biointensive Integrated Pest Management strategies and can be classified based on their feeding habits as Sucking, fruit and foliage feeders.

Insect pests	Scientific name	Family	Order
Fruits and foliage feeders			
Tobacco Caterpillar	*Spodoptera litura*	Noctuidae	Lepidoptera
Fruit borer	*Helicoverpa armigera*	Noctuidae	Lepidoptera
Cut worm	*Agrotis ipsilon*	Noctuidae	Lepidoptera
Sucking pests			
Chilli thrips	*Scirtothrips dorsalis*	Thripidae	Thysanoptera
Broad mite/yellow mite	*Polyphagotarsonemus latus*	Tarsonemidae	Acarina
Green peach Aphid	*Myzus persicae*	Aphididae	Hemiptera

1. Tobacco caterpillar

Systemic Position

Kingdom	:	Animalia
Phylum	:	Arthropoda
Class	:	Insecta
Order	:	Lepidoptera
Family	:	Noctuidae
Super family	:	Noctuoidea
Genus	:	*Spodoptera*
Species	:	*litura*
Binomial Name	:	*Spodoptera litura* (Fabricus)

Host range

Spodoptera litura is a notorious leaf feeding insect pest of more than one hundred plants around the Asia-Pacific region. Host plant survey for this insect pest revealed 27 plant species as host plants of *S. litura* belonging to 25 genera of 14 families including cultivated crops, vegetables, weeds, fruits and ornamental plants[24]. Major host plants on which it thrived for maximum period were Cotton (*Gossypium hirsutum* L.)[25], Castor (*Ricinus communis* L.), Cabbage (*Brassica oleracea* var. *botrytis* L.), Taro (*Colocasia esculenta* L.), Pigweed (*Trianthema portulacastrum* L.) and Sesban (*Sesbania sesban* L.)[23], tobacco (*Nicotiana tabacum*), Chinese cabbage (*Brassica rapa* subsp. pekinensis), cowpea (*Vigna unguiculata*) and sweet potato (*Ipomoea batatas*)[22], etc. There are 112 reported host plant of *S. litura*[15] out of which 60 host plants had been reported in India[17].

Distribution

The common cutworm, *Spodoptera litura*, is a pest of many crops across globe throughout tropical and temperate Asia, Australasia and the Pacific Islands and has become a major pest of soybean (*Glycine max*) throughout its Indian range[18-19]. It occurs mostly in the area having temperature range of 15-36°C[18]. In India, the incidence of this pest occurs in nearly all states but the predominant infestation is reported in Orissa, Karnataka, Maharashtra, Rajasthan, Andhra Pradesh, Tamil Nadu, Uttar Pradesh, Madhya Pradesh, etc.[21-22].

Symptoms of damage

A single larva per square metre is reported to cause average pod yield loss of 27.3% in groundnut through damage to various plant parts like leaves, flowers and pods[20]. The larva has vigorous eating patterns mostly included scraping of leaves, many times leaving the leaves only with its mid rib which directly decrease down the area of photosynthesis and hence reduce down the yielding potential of crop[26]. The moth's effects are quite disastrous, destroying economically important agricultural crops and decreasing yield in some plants completely[27-28].

Seasonal incidence

In *Kharif* season, the pest remains active from end of July or mid of August to October coinciding with warm and humid climate and peak reproductive phases of soybean, causing 26-29% yield losses[33]. In Rabi/Summer Season the pest incidence starts from 5th meteorological week onwards means from the last week of January and its peak period is reported in 11-12 meteorological week *i.e.* last fortnight of March and remain up to harvesting period[41].

Biology

Moths are active at night. The body of moth is stout with forewings pale grey to dark brown in colour with wavy white crisscross markings. Hind wings are whitish in colour with brown patches along wing margin. Female laid approx 300 eggs in clusters and its covered over by brown hairs. They hatch in about 3-5 days. Caterpillars looks like velvety and black in colour with yellowish green dorsal stripes and lateral white bands with incomplete ring like dark band on anterior and posterior end of the body. It passes through 6 instars. Larval stage lasts 15-30 days. Pupation takes place inside the soil. Papal stage lasts 8-15 days. Pest breeds throughout the year. Adults live for 7-10 days. Total life cycle takes 32-60 days. There are eight generations in a year.

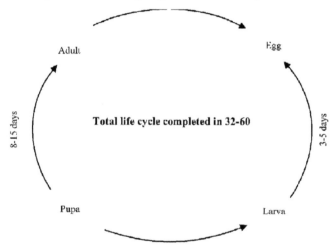

Fig. Diagrammatic representation of life cycle of Tobacco caterpillar, *Spodoptera litura* (Fabricius)

2. Fruit borer

Systemic Position

Kingdom	:	Animalia
Phylum	:	Arthropoda
Class	:	Insecta
Order	:	Lepidoptera
Family	:	Noctuidae
Super family	:	*Noctuoidea*
Genus	:	*Helicoverpa*
Species	:	*armigera*
Binomial Name	:	*Helicoverpa armigera* (Hübner)

Host range

Gram pod borer, *Helicoverpa armigera* (Hübner) is a polyphagous pest[45,46,48] of many crops in differentparts of the world and is reported to infest more than 60 plant species of 47 families[29,33,43]. Among whichsome of primarily including hosts are cotton (*Gossypium hirsutum* L.), corn (*Zea mays* L.), tomato (*Lycopersicon esculentum* Mill), hot pepper (*Capsicum frutescens* L.), tobacco (*Nicotiana tobacum* L.), common bean (*Phaseolus vulgaris* L.), sunflower (*Helianthus annuus* L.)[30], pigeon pea (*Cajanus cajan*), chickpea (*Cicer arietinum*), rice (*Oryza sativa*), sorghum (*Sorghum bicolor*) cowpea (*Vigna unguiculata*), Chilli (*Capsicum annuum*), soybean (*Glycine max*)[32], etc. It is reported that, many crop plants produces the particular stimulus to attract the insect-pest, natural enemies and other biota and form its own ecology which is mostly generalized with the sensing ability of *Helicoverpa armigera*[49,50,51].

Distribution

This lepidopteran pest is distributed all over the world *i.e.* eastwards from southern Europe and Africa through the Indian subcontinent to Southeast Asia, and spread up to China, Japan, Australia and the Pacific Islands[31,37]. It is mostly occurred from all parts of India like Madhya Pradesh, Assam, Orissa, West Bengal, Delhi, Haryana, Himachal Pradesh, Rajasthan, Punjab, Uttar Pradesh, Gujarat, Maharashtra, Andhra Pradesh, Karnataka, Tamil Nadu, etc.[45,51].

Symptoms of damage

Young larva feeds on the leaves, shoots, buds and some time and then attacks fruits. Internal tissues are eaten severely and completely hollowed out. During feeding the caterpillar the position of head inside and leaving the rest of the body outside. Bored fruits look like round holes.

Seasonal incidence

Pod borer starts activity on green gram, summer vegetables and maize and continues their generation by Aug-Sept months and synchronizing with main crop. In Rabi Season, the incidence of it starts from the December month only and remain up to the harvest of crop (March-April) and peak period of incidence found in the last fortnight of March[42,50].

Biology

Fore wing of moth is brownish or grayish in colour with a dark cross band near outer margin and dark spots near costal margins. Female laid spherical, yellowish eggs singly on tender parts and buds of plants. The incubation period varied from 2-4 days. The Caterpillars initially brown in colour and later turn into greenish with darker broken lines along the side of the body. The larval period varied from 18-25 days. While body covered with radiating hairs. The caterpillar pupates in the soil in an earthen cell and emerges after 7-15 days.

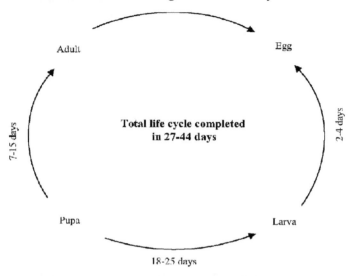

Fig. Diagrammatic representation of life cycle of Chilli Fruit Borer, *Helicoverpa armigera* Hubn

3. Cut worm

Systemic Position

Kingdom	:	Animalia
Phylum	:	Arthropoda
Subphylum	:	Uniramia
Class	:	Insecta
Order	:	Lepidoptera
Family	:	Noctuidae
Genus	:	Agrotis
Species	:	*ipsilon*
Binomial Name	:	*Agrotis ipsilon* (Hufnagel)

Host range

It has a wide range of host that includes grasses, weeds, and agricultural crops such as okra, onion, celery, groundnut, sugerbeet, rape, mustard, etc., apart from chilli[53,54,]. It is a polyphagous pest and mostly feeds on solanacious crops besides feeding on many other crop families[56,59] and responsible for causing huge losses when finds favorable conditions.

Distribution

Among the cutworm complex, *Agrotis ipsilon* are near-cosmopolitan noctuid moths most widely distributedin Asian and North American continent such as Afghanistan, India, Bangladesh, Cambodia, China, Egypt, etc[58]. In India it is distributed in states such as Assam, Bihar, Andhra Pradesh, Gujarat, Haryana, Himachal Pradesh, etc[59].

Symptoms of damage

Larva is the damaging stage. The first instars feeds on the tender leaves by making shot holes and the later instars cut the new seedlings during night which causes wilting of the plant and loss of the vigour[59]. During next plant growth stages, the later instars damages by tunneling into the plants. They hide during the day while feeds during the night. The shot holes, tunnels, leaf cuttings are the characteristic symptoms inflicted by the cutworm[63].

Seasonal incidence

The cutworms are active from March to September in tropics while March to October in subtropical and temperate areas[60]. The cut worms generally attainits peak population during May month and in fall of September and October months particularly in tropical regions[62].

Biology

The number of generation of this moth differs from one climatic regions to another climatic region with respect to whether parameter of particular area. It is active in tropical areas than in temperate one and hence, complete 2-4 generations there than 1-2 generation in temperate one. So, on an average one life cycle lasts for 35-55 days[53].

The female moths oviposit on the leaf or in the fence row (pasture) debris but not on bare soil, singly or in cluster of 1200-1900 eggs, with an incubation period of 3-6 days[55]. The egg is initial white colour which later changes to brown one with time and spherical one with 35-40 ribs that radiates from the apex having[56]

As the egg hatches the cannibalistic larva of light gray colour which later changes to dark gray and black having body size between 3.5 to 55 mm comes out. There are 5-9 larval instars depending on temperature (optimum temp. is 27°C) and diet.After 4th instar the larva becomes sensitive to day light and hence enters the soil during day while return to the surface during night. The larva pupates in 25-35 days[57].

The pupa is of dark brown colour and the pupation usually occurs in the soil at the depth of 2-12 mm, with a pupal period of about 12-20 days.

Adults have dark brown coloured forewings having light irregular and black dash marks with a span of 40-55 mm[59].

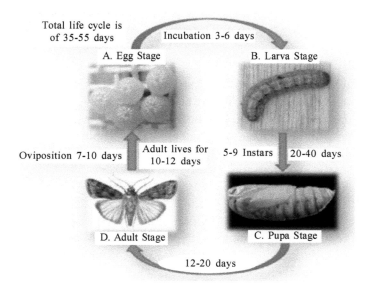

Fig. Diagrammatic representation of life cycle of Cut Warm, *Agrotis ipsilion* (Fabricius)

4. Chilli thrips

Kingdom	:	Animalia
Phylum	:	Arthropoda
Class	:	Insecta
Order	:	Thysanoptera
Family	:	Thripidae
Subfamily	:	Thripinae
Genus	:	*Scirtothrips*
Species	:	*dorsalis*
Binomial Name	:	*Scirtothripsdorsalis* (Hood)

Host range

It has been reported that this pest infest more than 100 plant taxa among 40 families for its nutrients, shelter and other requirements[67]. Plants belong to family "Fabaceae" are considered as its primary host[68]. In southest Asia and Indian subcontinent, *S. dorsalis* is considered as one of the most serious threats to chilli crop[69-70].There are many potential economic hosts reported for this pest like banana, bean, cashew, castor, citrus, cocoa, corn, cotton, egg plant, grapes, kiwi, litchi, longan, mango, melon, peanut, pepper, poplar, rose, strawberry, sweet potato, tea, tobacco, tomato, and wild yams (*Dioscorea* spp.)[73,74].

Distribution

It is a cosmopolitan pest reported its existence from various parts of the world like Australia, Africa, America, Asia, Europe, Thailand, Japan, China, etc[73]. In many parts it is result of invasion and encroachment via various means[74].

Symptoms of damage

The damaging stages of *S. dorsalis* both nymphs and adults which likes to feed on the tender and young parts only like soft meristem, leaves, flower buds and young fruits of the host plant by sucking sap from it which results in distortion and upward curling of leaves, discoloration of buds, flowers and young fruits[70]. The *S. dorsalis* possess chewing and lacerating type of mouth parts which can damage the host plant by extracting the contents of individual epidermal cells, leading to the necrosis and crinkling of tissue[73]. The color of damaged tissue changes from silvery to brown or black. The appearance of discolored or disfigured plant parts suggests the presence of *S. dorsalis*. Damage caused by *S. dorsalis* responsible for yield loss in a range between 61 to 74%[74].

Seasonal incidence

In India, the activity and appearance of thrips on chilli crop during Kharif season starts from fourth week of July and its peak population is reported during last week of August[61]. In hilly areas, the incidence of thrips has been recorded in the month of September (37th Standard Meteorological Week) and remain continued up to the last week of December (52nd SMW)[64,65]. The peak population of thrips in hilly region is recorded in the 40-44th SMW[66].

Biology

The chilli thrips is recorded as opportunistic species based on their life cycle pattern[68]. The stages of the life cycle of *S. dorsalis* include the egg, first and second instar larva, prepupa, pupa and adult[70]. The mating does not results in fertilization of all the eggs and unfertilized eggs produce males while fertilized eggs produce females[71]. As most of the eggs got fertilized by mating so the sex ratio is in favor of female progeny. Gravid females after mating lay single eggs inside the plant tissue (above the soil surface) and eggs hatch between 5-8 days depending upon environmental conditions[73]. As egg hatched, larva comes out and gathers to mid-veins and damaged portion of leaf tissues. There are two larval instars are reported in chilli thrips and both laval and adult stages are damaging one[94,95]. Larval stages complete in 8-10 days and pupation takes place in soil litter. Pupal stage comprised of two false pupal stages which are inactive and non-feeding one which lasts for about 2.6-3.3 days[74].

Adult females are approx. 1 mm in length and have pre oviposition period of approx. 1-2 days and each female lays approx.40 eggs. Total duration of life cycle varies between 30-45 days[75].

5. Broad mite/ yellow mite

Systemic Position

Kingdom	:	Animalia
Phylum	:	Arthropoda
Class	:	Arachnida
Subclass	:	Acari
Order	:	Trombidiformes
Family	:	Tarsonemidae
Genus	:	*Polyphagotarsonemus*
Species	:	*latus*
Binomial Name	:	*Polyphagotarsonemus latus* (Banks)

Host range

It belongs to the phytophagous mite family Tarsonemidae[77]. This species has a large host range world-wide and found on species belonging to more than 60 different plant families[79]. Jute, cucumber, cotton, grapevine, soybean, lemon, powpow, chilli, citrus, tea, etc. act as a main host for it and other host includes many plant and crop species belonging to different families like, pepper, mandarin, coffee, balsom, jatropa, mango, avocado, beans, etc[82-84].

Distribution

It is widely distributed throughout the tropics and sub tropics and is known by a number of common names[77]. It is reported to be present in all parts of world like Asia, Europe, North and South America, Africa, and Australia[79,82].

Symptoms of damage

Mites are usually seen on the newest leaves and small fruit and they causes terminal leaves and flower buds to become malformed by intoxication of plant parts which leads to formation of twisted, hardened and distortedness in topical growing parts[79]. Downward curling and crinkling of leaves giving an inverted boat shaped appearance, elongation of petiole *i.e.* rat tail with their coppery appearance is the characteristic feature of mite attack[80]. Heavy incidence results in stunted plant growth and abortion of bloom which indirectly reduced down the photosynthetic efficiency of plant which results in reduction of yield and productivity[81].

Seasonal incidence

During *kharif* season, the incidence of mites is observed in field from October to November which is significantly correlated with the presence of dry humid weather[77]. The infestation of mite commenced in the last week of August (35 SMW) and observed in the crop till third week of December (51 SMW)[11-13]. The highest mite population generally recorded in the first week of October (40 SMW)[10]. The incidence of mites is mostly found to be positively correlated with temperature and relative humidity but negatively correlated with the rainfall[85].

Biology

The broad mite is usually observed and feeds on the lower surfaces of apical leaves and in flowers, and there only its will lay elongate-oval shaped eggs. The eggs are laid in such way that, their upper surfaces covered with rows of whitish hemispherical projections, whereas their ventral, flat bases are firmly appressed to the plant substrate[76]. At optimum temperature the hatching of eggs takes place in 2-3 days. Emerging larvae are not very active; they feed for 1 day only and molt into pharate females or quiescent nymphs[77]. Pharate females (often termed 'pupae' or 'quiescent nymphs') remain for another day within their larval skins, from which they subsequently emerge. Total development thus requires around 5 days at $25^{\circ}C$. Males appear first and locate the pharate females for future mating purpose and carry them on their posterior dorsal side with the help of genital papillae towards plants apical parts[78]. Mating occurs for 15-120 s immediately after emergence of female while the genders face in opposite directions. Unmated females may also copulate later during their lives, but it is not known whether they mate more than once. It is observed that males are very active particularly during warm season and lives for a week while female are comparatively sluggish and they lives for some another days than males[76]. The fairly mild and humid weather is considered as an optimum condition for the reproduction of broad mite with population growth being highest at $25^{\circ}C$

andaverage fecundity observed is usually around 40 eggs/mated female (range: 30-76).

The usual adult male:female ratio is 1:4, although it may fluctuate. Under dry conditions many female immatures do not survive to adulthood, and males become more abundant[78]. Like other tarsonemids, *P. latus* reproduces by arrhenotoky, unmated females producing only males. Sons of virgin females can mate with their mothers, which may then produce female offspring, thus assuring the survival of broad mite populations. Dispersal is effected through various means[79]. Within infested plants, *P. latus* moves upwards by male carriage of pharate females; in this way males 'choose' the female's oviposition sites[83]. Mites reach uninfested hosts with winds, while phoretic on plant-feeding insects, and by human transfer of infested plants. Finally, it has been also reported that active mite individuals 'crossed' (presumably by walking) an 18-inch space between plants[76].

6. Green peach aphid

Systemic Position

Kingdom	:	Animalia
Phylum	:	Arthropoda
Class	:	Insecta
Order	:	Hemiptera
Family	:	Aphididae
Genus	:	Myzus
Species	:	*persicae*
Binomial Name	:	*Myzus persicae* (Sulzer)

Host range

Myzus persicae is well adapted to many host plants belonging to more than thirty different families and able to transmit over 100 viral diseases of plants of different species[87]. The notable host crops for this species are citrus, tobacco, brassica, sugar cane, sugar beet, beans, etc[89-92].

Distribution

It is a cosmopolitan and polyphagous pest and so recorded from many parts of world[87]. Except Arctic and Antarctic region it is mostly present in all continents like Europe, Asia, Australia, Africa, and North and South America[90-91].

Symptoms of damage

The damage caused by *M.persicae* can be seen as a chlorosis and necrosis, wilting and stunting, distortion and defoliation of crop leaves by its extensive feeding which leads to reduction in rate of photosynthesis and transpiration and increases the rate of respiration[90].

Aphids generally produces large quantities of sticky exudates known as honeydew while feeding on leaves of crop plant which shelter and inspires the growth of fungus called black sooty mold which directly reduce down the photosynthetic ability and indirectly adversely affect the yield of crop produce[89].

This pest is of even more importance when it act as a vector and responsible for carrying the inoculum of many plant viruses such as pepper mottle virus (PMV), potato virus Y (PVY), tobacco etch virus (TEV) and cucumber mosaic virus (CMV)[91].

Seasonal incidence

In *Kharif* season, the incidence of aphid on chilli starts with the onset of mansoon and their population starts increasing in the month of September and October[10]. It is observed that the incidence of aphid in field is positively correlated with the relative humidity and negatively correlated with the rainfall[11].

In *Rabi* season, the population of aphids mostly appeared in initial growth stages of crop plant and starts declining with onset of flowering and fruit setting stage[12]. There is a negative correlation exists between temperature and aphid population in field *i.e.* as temperature starts increasing than optimum level the aphid populations starts decreasing[13].

Biology

Green Peach Aphid, is aheterocious and holocyclic (host alternating, with sexual reproduction during part of life-cycle) mostly in temperate regions but anholocyclic in tropics and some sub-tropics regions where moderate climatic conditions exists and support the winter generations of the pest with exceptions: for example, it is reported apterous oviparous females of *M. persicae* for the first time from India, collected on *Prunus persica*[87].

In case of host alternating populations, the Winged Fundatrigeniae migrates to the summer host also known as secondary host which includes crop plants and some herbacious shrubs, weeds, etc[88]. and there they produce wingless Verginoparae by thelitokous parthenogenesis. Wingless Virginoparae makes a firm hold and attachment with summer host plant by adopting the physiochemical conditions of new host and start the production of new generations *i.e.* winged virginoparae and vice versa for about 30-35 generations if an optimum condition prevails[89]. As the day length and temperature starts declining than the optimum one and relative humidity and cold waves in atmosphere increases, Wingless Virginoparae starts the production of winged Gynoparae which acts as an Autumn Migrants and starts to migrates toward the primary host plants which are mostly a woody trees. As the Winged Gynoparae reaches to primary host and relatives they established well but starts the production of wingless oviparae (mating females) or sexual cycle only on *Prunus persica*. Consequently, the wingless virginoparae starts the production of Winged Males nearly one month after the production of gynoparae which migrates independently towards the primary host to copulate with the Wingless Oviparae which by then has become adult or sexually mature[87].

It has been observed that males borne on summer host got attracted to female by the pheromone released by female from winter host or they are also attracted to the semiochemicals released by winter host plants[91].

The mating of Oviparae and Male results in laying of fertilized eggs (4-13 eggs) by Oviparae in cracks and crevices or terminal portions of winter host or primary hosts. On an average one Prunus tree can behold maximum 20,000 eggs and generally 4000 eggs normally found on a tree which varies with the height and nature of tree. Aphids generally overwinter in diapause stage *i.e.* eggs which require cold and chilling temperature (>-46^{0}C) for its development. The hatching of eggs is usually associated with swelling of flower bud of *Prunus* which provides food for new born Fundatrix and due to many conditions high mortality in new born *i.e.* Fundatrix may even occure[90].

Adult Fundatrix undergoes the litokous parthenogenesis which results in production of winged and wingless Fundatrigeniae viviparaosly which feeds upon flower buds and soft tissues of the peach tree. Wingless Fundatrigeniae continue to be produces from Fundatrix unless the nutritional quality of *Prunus* dose not decline, and as it starts declining the production of winged Fundatrigeniae also begin which is having a migratory nature[93]. Hence, winged Fundatrigeniae also called as a Spring Migrants which starts migration towards secondary summer hosts[91].

Biointensive integrated insect-pest management

The biointensive pest management deals with use of naturally obtain products for example from plants, animalor their derivatives either in direct or indirect way to monitor and regulate the insect pest management in the field. There are many organisms mostly multicellular or their derivatives which has potential to act as a natural enemy of one or more pests in field condition. Following has given the list of different natural enemies with respect to particular pest of Chili crop.

1. Tobacco caterpillar[37-39,96-101,106,108]

Egg Parasitoid

   Telenomus nawai (Hymenoptera: Scelionidae)

   Trichogramma ostriniae (Hymenoptera: Trichogrammatidae)

   Trichogramma chilonis

Larval Parasitoid

   Campoletis chlorideae   *Diadegma argenteopilosa*

   Eriborus argenteopilosus   *Apanteles* colemani

   Bracon brevicornis   *Campoletis chlorideae*

   Cotesia sesamiae   *Microgaster tuberculifera*

   Microplitis pallidipes

Pupal Parasitoid

   Eurytoma syleptae   *Winthemia*

Predators

   Rhynocoris marginatus   *Rhynocoris kumara*

   Eocanthecona furcellata

Entomopathogenic Fungi

   Nomuraea rileyi   *Metarhizium anisopliae*

   Beauveria bassiana   *Aspergillus ochraceus*

Entomopathogenic Nematodes

   Ovomermis albicans   *Hexamermis* sp.

   Pentatomermis sp.   *Steinernema* sp.

Entomopathogenic Microorganism

   Nosema carpocapsae (**Protozoa**)

   Bacillus thuringiensis Kurstaki (**Bacterial Formulation**)

   Cytoplasmic polyhedrosis virus (CPV) (**Virus Formulation**)

   Granulosis virus

   Spodoptera litura Nucleopolyhedrosis virus (SINPV)

Botanicals effective against Tobacco Caterpillar, Spodoptera litura

Neem oil, Karanj oil, Azadirachtin and karanjin@ 0.5 L ha⁻¹

Neem Seed Kernel Extract (NSKE)

Gokhru, *Pedalium murex* (root)

Common lantana, *Lantana camara* (leaves)

Gurmar,*Gymnema sylestre* (Leaves)

2. Fruit borer[14,33-40,42,51,104-106]

Egg Parasitoid   : *Telenomus   remus* (Hymenoptera: Scelionidae)

Trichogramma achaeae (Hymenoptera :Tricho- grammatidae)

Trichogramma chilonis, Trichogramma minutum Trichogramma bactriana, Trichogrammpretiosum

Copidosoma koehleri (Hymenoptera: Encyrtidae)

Microchelonus blackburni (Hymenoptera: Braco-nidae)

Larval Parasitoid

Eriborus argenteopilosus	*Apanteles diparopsidis*
Bracon kirkpatricki	*Campoletis chlorideae*
Cotesia kazak	*Microchelonus blackburni*
Microplitis rufiventris	*Winthemia lateralis*
Lespesia archippivora	*Glabromicroplitis croceipes*
Exorista xanthaspis	*Eucelatoria bryani*
Cotesia marginiventris	*Sturmiopsis inferens*

Pupal Parasitoid

Carcelia illota	*Eurytoma syleptae*
Heteropelma scaposum	

Predators

Acanthaspis pedestris	*Acridotheres tristis*
Chrysoperla carnea	*Cheilomenes sexmaculata*
Eocanthecona furcellata	*Coccinella septempunctata*
Harmonia axyridis	*Hippodamia variegate*
Mallada boninensis	*Euborellia pallipes*
Orius albidipennis	*Orius minutes*
Nabis capsiformis	*Nabis palifer*
Xysticus croceus	*Rhynocoris marginatus*
Eocanthecona furcellata	*Rhynocoris kumara*

Entomopathogenic Fungi

Nomuraea rileyi	*Metarhizium anisopliae*
Beauveria bassiana	

Entomopathogenic Nematodes

Steinernema feltiae	*Heterohabditis bacteriophora*
Steinernema carpocapsae	*Heterohabditis indica*

Entomopathogenic Microorganism

Vairimorpha necatrix (**Protozoa**)

Bacillus thuringiensis Kurstaki (**Bacterial Formulation**)

Serratia marcescens

Cytoplasmic polyhedrosis virus (CPV) (**Virus Formulation**)

Granulosis virus

Helicoverpa armigera Nucleopolyhedrosis virus (HaNPV)

Botanicals effective against Fruit Borer, *Helicoverpa armigera*:

Neem oil, Karanj oil, Azadirachtin and karanjin@ 0.5 L ha[-1]

Neem Seed Kernel Extract (NSKE)

Gokhru, *Pedalium murex* (root)

Common lantana, *Lantana camara* (leaves)

Gurmar,*Gymnema sylestre* (Leaves)

3. Cut worm[37-39,59]

Egg Parasitoid

Telenomus remus (Hymenoptera:Scelionidae) *Trichogramma dendrolimi* (Hymenoptera:Trichogrammatidae) *Trichogramma evanescens.*

Larval Parasitoid

Chelonus inanitus	*Apanteles bourquini*
Campoletis flavicincta	*Campoletis chlorideae*
Cotesia kazak	*Zele nigricornis*
Microplitis rufiventris	*Microplitis similis*
Gonia bimaculata	*Periscepsia carbonaria*
Exorista xanthaspis	*Exorista larvarum*
Siphona collini	*Siphona cristata*

Pupal Parasitoid

Archytas marmoratus	*Netelia fuscicornis*

Predators

Abacidus permundus	*Pterostichus chalcites*
Chrysoperla carnea	*Cyclotrachelus sodalist*
Zelus tetracanthus	*Stelopolybia pallipes*

Entomopathogenic Fungi

Nomuraea rileyi
Beauveria bassiana

Entomopathogenic Nematodes

Steinernema feltiae	*Noctuidonema guyanense*
Heterorhabditis bacteriophora	*Hexamermis heterocephalis*

Entomopathogenic Microorganism

Vairimorpha necatrix (**Protozoa**)
Bacillus thuringiensisKurstaki (**Bacterial Formulation**)
Serratia marcescens
Granulosis virus (**Virus Formulation**)
Nucleopolyhedrosis virus (NPV)

Botanicals effective against Cut Worm, *Agrotis ipsilon*:

Neem oil, Karanj oil, Azadirachtin and karanjin@ 0.5 L ha[-1]

Neem Seed Kernel Extract (NSKE)

Gokhru, *Pedalium murex* (root)

Common lantana, *Lantana camara* (leaves)

Gurmar, *Gymnema sylestre* (Leaves)

4. Chilli thrips[11-14,37-39,74,110,112]

Predators

Aduncothrips asiaticus	*Carayonocoris indicus*
Franklinothrips megalops	*Geocoris ochropterus*
Mymarothrips garuda	*Scolothrips indicus*

Botanicals effective against Chilli Thrips, *Scirtothrips dorsalis*

Neem Seed Kernel Extract (NSKE)	Azadirachtin 0.0075%

Entomopathogenic Microorganism

Bacillus thuringiensis Kurstaki **(Bacterial Formulation)**

Heterorhabditis indica (2000IJs/ml) **(EPN Formulation)**

Beauveria bassiana, Metarhizium brunneum,Isaria fumosorosea **(EPF)**

5. Broad mite/ yellow mite[11-14,37-39,107,109]

Predators

Amblyseius swirskii	*Amblyseius victoriensis*
Euseius ovalis	*Metaseiulus occidentalis*
Typhlodromus porresi	*Neoseiulus californicus*

Entomopathogenic Fungi

Hirsutella thompsonii

Beauveria bassiana

Botanicals effective against Broad mite/ yellow mite

Acorus calamus (L)	*Xanthium strumarium* (L)
Polygonum hydropiper (L)	*Clerodendron infortunatum* (Gaertn)

Azadirachtin 3% WSP   Pongam kernel aqueous extract (PKAE)
Garlic aqueous extract (GAE) *Lippia nodiflora*
Cassava roots

Entomopathogenic Microorganism

Liuyangmycin (an antibiotic preparation from *Streptomyces griseolus*)

6. Green peach aphid[11,12,14,37-39,87,102,103]

Parasitoids

Trioxys similis   *Trioxys angelicae*
Aphidius matricariae   *Aphelinus semiflavus*
Aphidius colemani   *Aphidius gifuensis*
Zele chlorophthalma   *Aphelinus mali*
Diaeretiella rapae   *Ephedrus californicus*

Entomopathogenic Fungi:

Lecanicillium lecanii   *Erynia neoaphidis*
Beauveria bassiana   *Verticillium lecanii*
Chromaphidis spp   *Entomophaga chromaphidis*
Conidiobolus obscurus   *Pandora neoaphidis*
Fusarium pallidoroseum   *Verticillium lamellicola*

Entomopathogenic Predators

Coccinella septempunctata *Chilomese sexmaculata*
Eocanthecona furcellata   *Common myna bird (Acridotheres tristis)*
Episyrphus balteatus   *Ischiodon scutellaris*
Orius insidiosus   *Micraspis discolor*
Metasyrphus corollae   *Scaeva pyrastri*
Scymnus posticalis   *Scymnus bicolor*

Entomopathogenic Microorganism

Bacillus thuringiensis Kurstaki (**Bacterial Formulation**)

Bacillus thuringiensis

Botanicals effective against Green Peach Aphid, *Myzus persicae*

Azadirachtin 1.5 EC (0.0025%)

Crofton weed, *Eupatorium adenophorum* based formulation

Common Mugwort, *Artemisia vulgaris*

Sweet flag, *Acorus calamus*

Billygoat-weed, *Ageratum conyzoides*

References

Abdel-Gawaad, A.A. and El-Shazli, A.Y. (1971). Studies on the common cutworm *Agrotis ypsilon* Rott. I. Life cycle and habits. *Zeitschrift fuer Angewandte Entomologie* 68: 409-412.

Ahmad, M., Ghaffar, A. and Rafiq, M. (2013). Host Plants of Leaf Worm, *Spodoptera Litura* (Fabricius) (Lepidoptera: *Noctuidae*) in Pakistan. *Asian J. Agri. Biol.*, 1(1): 23-28.

Ananthakrishnan, T.N. (1993). Bionomics of Thrips. *Annual Review of Entomology*, 38(1): 71-92.

Arthurs, S.P., Aristizábal, L.F. and Avery, P.B. (2013). Evaluation of entomopathogenic fungi against chilli thrips, *Scirtothrips dorsalis*. *Journal of Insect Science*, 13: 31.

Awasthi, A. and Awasthi, S. (2018). Screening of some plant extracts against polyphagous pest *Spodoptera litura* (Lepidoptera: Noctuidae). *International Journal of Zoology Studies*, 3(1): 173-176.

Bajpai, N.K. and Sehgal, V.K., (1994). Effect of neem products, nicotine and karanj on survival and biology of pod borer, *Helicoverpa armigera* Hubn. of chickpea. *Proc. II AZRA Conf. on Recent Trends in Plant, Animal and Human Pest Management: Impact on Environment*. Madras Christian College, Madras, 27-29 pp 48.

Bijaya, P., Subharani, S. and Singh, T.K. (2005). Bioefficacy of certain botanical insecticides against green peach aphid, *Myzus persicae* (Sulzer) on cabbage. Pestology.

Blackman, R.L. (1974). Life-cycle variation of *Myzus persicae* (Sulz.) (Hom., Aphididae) in different parts of the world, in relation to genotype and environment. *Bulletin of Entomological Research*, 63(04), 595. doi:10.1017/s0007485300047830

Brown, E.S. and Dewhurst, C.F. (1975). The genus *Spodoptera* (Lepidoptera, Noctuidae) in Africa and the Near East. *Bulletin of Entomological Research*, 65(2): 221-262.

Busching, M.K. and Turpin, F.T. (1976). Oviposition preferences of black cutworm moths among various crop plants, weeds, and plant debris. *Journal of Economic Entomology*, 69: 587-590.

Busching, M.K. and Turpin, F.T. (1977). Survival and Development of Black Cutworm (*Agrotis ipsilon*) Larvae on Various Species of Crop Plants and Weeds. *Environmental Entomology*, 6(1): 63–65. doi:10.1093/ee/6.1.63.

Chandrayudu, E., Murali Krishna, T., Sudheer, M.J., Sudhakar, P. and Vemana, K. (2015). Bio-efficacy of certain botanicals and bio-pesticides against tobacco caterpillar, Spodoptera litura Fab. in rabi groundnut. *Journal of Biological Control,* 29(3): 131-133.

Ciancio, A. and Mukarji, K.G. (2007). General Concepts in Integrated Pest and Disease Management. Springer, Dordrecht, The Netherlands.

Davis, J.A., Radcliffe, E.B., and Ragsdale, D.W. (2006). Effects of High and Fluctuating Temperatures on *Myzus persicae* (Hemiptera: *Aphididae*). *Environmental Entomology*, 35(6): 1461–1468. doi:10.1093/ee/35.6.1461

Dubey, N.K., Shukla, R., Kumar, A., Singh, P. and Prakash, B. (2011). Global Scenario on the application of Natural products in integrated pest management. *In: Natural Products in Plant Pest Management,* Dubey, N.K., CABI, Oxfordshire, UK., pp 1-20.

EPPO, (2014). PQR database. Paris, France: European and Mediterranean Plant Protection Organization. http://www.eppo.int/DATABASES/pqr/pqr.htm

European Food Safety Authority (EFSA), Schrader, G., Camilleri, M., Diakaki, M. and Vos, S. (2019). Pest survey card on *Scirtothrips aurantii*, *Scirtothrips citri* and *Scirtothrips dorsalis*. EFSA supporting publication 2019: EN-1564. 21 pp. doi:10.2903/sp.efsa.2019.EN-1564

Fand, B.B., Sul, N.T., Bal, S.K. and Minhas, P.S. (2015). Temperature Impacts the Development and Survival of Common Cutworm (*Spodoptera litura*): Simulation and Visualization of Potential Population Growth in India under Warmer Temperatures through Life Cycle Modelling and Spatial Mapping. PLoS ONE, 10(4): e0124682.

Fathipour, Y. and Naseri, B. (2011). Soybean Cultivars Affecting Performance of *Helicoverpa armigera*(Lepidoptera: Noctuidae). In: Ng TB. (ed.) Soybean-Biochemistry, Chemistry and Physiology. Rijeka: InTech; 599-630.

Fathipour, Y. and Sedaratian, A. (2013). Integrated Management of *Helicoverpa armigera* in Soybean Cropping Systems. IntechOpen. DOI: 10.5772/54522.

Gayatri Priya, N., Ojha, A., Kajla, M.K., Raj, A. and Rajagopal, R. (2012). Host Plant Induced Variation in Gut Bacteria of *Helicoverpa armigera*. PLoS ONE 7(1): e30768. doi:10.1371/journal.pone.0030768

Geetha, R. and Selvarani, K. (2017). A study of chilli production and export from India. *IJARIIE*, 3(2): 205-210.

Gerson, U. (1992). Biology and control of the broad mite, *Polyphagotarsonemus latus* (Banks) (Acari: *Tarsonemidae*). *Experimental and Applied Acarology*, 13(3): 163-178.

Ghumare, S.S. and Mukherjee, S.N. (2003). Performance of *Spodoptera litura* Fabricius on different host plants: influence of nitrogen and total phenolics of plants and mid-gut esterase activity of the insect. *Indian J. Exp. Biol.*, 41(8): 895-9.

Giraddi, R.S., Abhilash, H.R., Mallapur, C.P., and Girish, V.P. (2018). Present status of chilli murda and its management-An overview. J. Farm Sci., 31(4): (359-368).

Glazer, I. (1997). Effects of infected insects on secondary invasion of steinernematid entomopathogenic nematodes. *Parasitology*, 114(6): 597-604.

Gomes, E.S., Santos, V. and Ávila, C.J. (2017). Biology and fertility life table of *Helicoverpa armigera* (Lepidoptera: *Noctuidae*) in different hosts. *Entomological Science*, 20: 419–426.

Gopal, G.V., Lakshmi, K.V., Babu, B.S. and Varma, P.K. (2018). Seasonal incidence of chilli thrips, *Scirtothrips dorsalis* Hood in relation to weather parameters. *Journal of Entomology and Zoology Studies*, 6(2): 466-471.

Greenberg, S.M., Sappington, T.W., Legaspi, B.C., Liu, T.-X. and Tamou, M.S. (2001). Feeding and Life History of *Spodoptera exigua* (Lepidoptera: Noctuidae) on Different Host Plants. *Arthropod Biology*, 94(4): 565-574.

Grinberg, M., Perl-Treves, R., Palevsky, E., Shomer, I. and Soroker, V. (2005). Interaction between cucumber plants and the broad mite, *Polyphago-tarsonemus latus*: from damage to defense gene expression. Entomologia Experimentalis et Applicata, 115(1), 135–144. doi:10.1111/j.1570-7458.2005.00275.x

Hardwick, D.F. (1965). The corn earworm complex. *Memoirs of the Entomological Society of Canada*, 40: 1-247. (Biology)

Harshita, A.P., Saikia, D.K., Anjumoni Deeve, Bora, L.C. and Phukan, S.N. (2018). Seasonal incidence of fruit borer, *Helicoverpa armigera* and its eco-friendly management in tomato, *Solanum lycopersicum*. *International Journal of Chemical Studies*, 6(5): 1482-1485.

Havanoor, R. and Rafee, C.M. (2018). Seasonal incidence of sucking pests of chilli (*Capsicum annum* L.) and their natural enemies. *Journal of Entomology and Zoology Studies*, 6(4): 1786-1789.

Hemati, S.A., Naseri, B., Ganbalani, G.N., Dastjerdi, H.R. and Golizadeh, A. (2012). Effect of Different Host Plants on Nutritional Indices of the Pod Borer, *Helicoverpa armigera*. *Journal of Insect Science*, 12(55): 1–15. doi:10.1673/031.012.5501

Horticultural Statistics at a Glance (2018). Horticulture Statistics Division, Department of Agriculture, Cooperation and Farmers' Welfare, Ministry of Agriculture and Farmers' Welfare, Government of India.

Hossen, S.M. (2015). A Statistical Analysis on Production of Chili and Its' Prospect in Bangladesh. Global Disclosure of Economics and Business, 4(1): 55-61.

http://entnemdept.ufl.edu/creatures/orn/broad_mite.htm

https://www.cabi.org/isc/datasheet/26757

https://www.cabi.org/isc/datasheet/26876

https://www.cabi.org/isc/datasheet/35642

https://www.cabi.org/isc/datasheet/3801

https://www.cabi.org/isc/datasheet/44520

https://www.cabi.org/isc/datasheet/49065

Jagdish, E.J. and Purnima, A.P. (2011). Evaluation of selective botanicals and entomopathogens against *Scirtothrips dorsalis* Hood under polyhouse conditions on rose. *Journal of Biopesticides*, 4 (1): 81-85.

Jagtap, P.P., Shingane, U.S. and Kulkarni, K.P. (2012). Economics of Chilli Production in India. *African Journal of Basic and Applied Sciences*, 4(5):161-164.

Jangra, M., Gulati, R. and Sonika (2017). Incidence of chilli mite, *Polyphagotarsonemus latus* (Banks) on chilli fruit parameters under field conditions. *Emer. Life Sci. Res.*, 3(2): 26-31.

Jenser, G. and Szénási Á. (2004) Review of the biology and vector capability of *Thrips tabaci* Lindeman (Thysanoptera: *Thripidae*). *Acta Phytopathologica et Entomologica Hungarica*, 39(1-3):14.

Jenser, G., Lipcsei, S., Szénási, Á. and Hudák, K. (2006). Host Range of the Arrhenotokous Populations of *Thrips Tabaci* (Thysanoptera: *Thripidae*). *Acta Phytopathologica et Entomologica Hungarica*, 41 (3–4), 297–303.

Jorwar, R.M., Sarap, S.M. and Chavan, V.U. (2018). Economics of production and marketing of chilli in Amravati district. *Journal of Pharmacognosy and Phytochemistry*, 7(2): 310-316.

Joshi, B.G., Sitaramaiah, S., Satyanarayana, S.V.V. and Ramaprasad, G. (1979). Note on natural enemies of *Spodoptera litura* (F.) and *Myzus persicae*

(Sulz.) on flue-cured tobacco in Andhra Pradesh. *Science and Culture*, 45(6): 251-252.

Katayama, J. and Sano, K. (1989). Damage analysis of red bean plants caused by the common cutworm, *Spodoptera litura* (Lepidoptera: *Noctuidae*). *Japanese Journal of Applied Entomology and Zoology*, 33(2): 57-62.

Khater, H.F. (2011). Ecosmart Biorational Insecticides: Alternative Insect Control Strategies. In: Insecticides, *Perveen, F. (Ed.)*. *InTech, Rijeka, Croatia*, ISBN 979-953-307-667-5.

Kousik, C.S., Shepard, B.M. Hassell, R., Levi, A. and Simmons, A.M. (2007). Potential Sources of Resistance to Broad Mites (*Polyphagotarsonemus latus*) in Watermelon Germplasm. *HortScience*, 42(7):1539–1544.

Kumar, R., Ahad, I., Rehman, S.A. and Dorjey, S. (2018). Impact of weather parameters on population dynamics of soil borne insect pests infesting oats (*Avena sativa* L.) in North Kashmir. *Journal of Entomology and Zoology Studies*, 6(3): 533-537.

Kumar, S. and Gavkare, O. (2015). Efficacy of chemical and botanical insecticides against greenpeach aphid, *Myzus persicae* on sweet pepper under protected environment. *Indian Journal of Plant Protection*, 43(4): 508-510.

Kumar, V., Kakkar, G., McKenzie, C.L., Seal, D.R. and Osborne, L.S. (2013). An Overview of Chilli Thrips, *Scirtothrips dorsalis* (Thysanoptera: Thripidae) Biology, Distribution and Management. Weed and Pest Control - Conventional and New Challenges. Sonia Soloneski and Marcelo Larramendy, IntechOpen, DOI: 10.5772/55045.

Latha, S. and Hunumanthraya, L. (2018). Integrated management of insect and mite pests of chilli under hill zone of Karnataka. *Journal of Entomology and Zoology Studies*, 6(2): 2770-2773.

Leibee, G.L., Kok-Yokomi, M.L., Aristizabal, L.F., Arthurs, S.P. and Morales-Reyes, C. (2015). Control of Chilli Thrips with Botanical Insecticides in Knock Out Rose. *Arthropod Management Tests*, 40(1):G7. doi:10.1093/amt/tsv183

Li, W., Teng, X., Zhang, H., Liu, T., Wang, Q., Yuan, G., and Guo, X. (2017). Comparative host selection responses of specialist (*Helicoverpa assulta*) and generalist (*Helicoverpa armigera*) moths in complex plant environments. PLOS ONE, 12(2), e0171948. doi:10.1371/ journal. pone. 0171948

Liu, Y., Fu, X., Mao, L., Xing, Z. and Wu, X. (2016). Host Plants Identification for Adult *Agrotis ipsilon*, a Long-Distance Migratory Insect. Int. J. Mol. Sci., 17, 851; doi:10.3390/ijms17060851.

Liu, Z., Li, D., Gong, P. and Wu, K. (2004). Life Table Studies of the Cotton Bollworm, *Helicoverpa armigera* (Hübner) (Lepidoptera: *Noctuidae*), on Different Host Plants, *Environmental Entomology*, 33(6): 1570–1576.

Liu, Z., Li, D., Gong, P. and Wu, K. (2004). Life Table Studies of the Cotton Bollworm, *Helicoverpa armigera* (Hübner) (Lepidoptera: *Noctuidae*), on Different Host Plants. *Environmental Entomology*, 33(6): 1570–1576. doi:10.1603/0046-225x-33.6.1570

Mandal, S.K., S.B. Sah and S.C. Gupta, (2007). Management of insect pests on okra with biopesticides and chemicals. *Ann. Plant Protection Sci.*, 15: 8791.

Mandi, N. and Senapati, A.K. (2009). Integration of chemical botanical and microbial insecticides for control of thrips, *Scirtothrips dorsalis* Hood infesting chilli. *The Journal of Plant Protection Sciences*, 1(1): 92-95.

Margaritopoulos, J.T., Kasprowicz, L., Malloch, G.L., and Fenton, B. (2009). Tracking the global dispersal of a cosmopolitan insect pest, the peach potato aphid. BMC Ecology, 9(1), 13. doi:10.1186/1472-6785-9-13

Margaritopoulos, J.T., Tsitsipis, J.A., Goudoudaki, S. and Blackman, R.L. (2002). Life cycle variation of *Myzus persicae* (Hemiptera: *Aphididae*) in Greece. *Bulletin of Entomological Research*, 92: 309–319.

Meena, R.K. and Tayde, A.R. (2017). Effect of Abiotic Factors on the Population Dynamics of Chilli Thrips, *Scirtothrips dorsalis* (HOOD) In Chilli, *Capsicum annum* L. Crop in Allahabad, India. *Int.J.Curr.Microbiol. App.Sci*, 6(6): 2184-2187.

Montasser, A.A., Taha, A.M., Hanafy, A.R.I. and Hassan, G.M. (2011). Biology and control of the broad mite, *Polyphagotarsonemus latus* (Banks, 1904) (Acari: *Tarsonemidae*). *IJESE*, 1: 26-34.

Natikar, P.K. and Balikai, R.A. (2015). Tobacco caterpillar, Spodoptera litura (fabricius): toxicity, ovicidal action, oviposition deterrent activity, ovipositional preference and its management. *Biochem. Cell. Arch.*, 15(2): 383-389.

Natikar, P.K. and Balikai, R.A. (2017). Present status on bio-ecology and management of tobacco caterpillar, *Spodoptera litura* (Fabricius) – An update. *International Journal of Plant Protection*, 10(1): 193-202.

Nikolakakis, N.N., Margaritopoulos, J.T. and Tsitsipis, J.A. (2003). Performance of *Myzus persicae* (Hemiptera: Aphididae) clones on different host-plants and their host preference. *Bulletin of Entomological Research*, 93: 235–242.

Onzo, A., Houedokoho, A.F. and Hanna, R. (2012). Potential of the predatory mite, *Amblyseius swirskii*to suppress the Broad Mite, *Polyphago-*

tarsonemus latus on the Gboma Eggplant, *Solanum macrocarpon*. *Journal of Insect Science*, 12(7), 1–11. doi:10.1673/031.012.0701

Padaliya, S., Thumar, R.K. and Timbadiya, B.G. Population dynamics of thrips, *Scirtothrips dorsalis* Hood infesting *Bt* cotton ecosystem in middle Gujarat. *Journal of Entomology and Zoology Studies*, 7(1): 420-422.

Palevsky, E., Soroker, V., Weintraub, P., Mansour, F., Abo-Moch, F. and Gerson, U. (2001). How species-specific is the phoretic relationship between the broad mite, *Polyphagotarsonemus latus* (Acari: *Tarsonemidae*), and its insect hosts? *Experimental and Applied Acarology*, 25(3): 217-224.

Palmer, J.M. and Mound, L.A. (1983). The *Scirtothrips* species of Australia and New Zealand (Thysanoptera: Thripidae). *Journal of Natural History*, 17(4): 507-518.

Patel, B.H., Koshiya, D.J. and Korat, D.M. (2009). Population dynamics of chilli thrips, *Scirtothrips dorsalis* Hood in relation to weather parameters. *Karnataka J. Agric. Sci.*, 22(1): 108-110.

Pathipati, V.L., Vijayalakshmi, T. and Naidu, L.N. (2014). Seasonal incidence of major insect pests of chilli in relation to weather parameters in Andhra Pradesh. Pest Management in Horticultural Ecosystems, 20(1): 36-40. (Thrips, mites, sl, Ha).

Patil, V.M., Patel, Z.P., Oak, P.S., Chauhan, R.C. and Kaneriya, P.B. (2018). Biology of fruit borer, *Helicoverpa armigera* (Hubner) in/on chilli fruits. *International Journal of Entomology Research*, 3(1): 06-12.

Pawar, S.S., Bharude, N.V., Sonone, S.S., Deshmukh, R.S., Raut, A.K. and Umarkar, A.R. (2011). Chillies as food, spice and medicine: A perspective. *International Journal of Pharmacy and Biological Sciences*, 1(3): 311-318. Mehta, I. (2017). *Chillies*– The Prime Spice – A History. IOSR *Journal of Humanities and Social Science*, 22(7)9: 32-36.

Perkins, L.E., Cribb, B.W., Hanan, J., Glaze, E., Beveridge, C. and Zulucki, M.P. (2008). Development and feeding behavior of Helicoverpa armigera (Hübner) (Lepidoptera: Noctuidae) on different sunflower genotypes under laboratory conditions, Arthropod-Plant Interactions, Springer, 2: 197. https://doi.org/10.1007/s11829-008-9047.

Prakash, A. and Rao, J. (2016). Biointensive Insect Pest Management,National Symposium on "New Horizons in Pest Management for Sustainable Goals" on 24-25 November, 2016 at OUAT, Bhubaneswar.

Rajput, V.S., Prajapati, B.G., Pareek, A. and Patel, P.S. (2017). Studies on Population Dynamics of Major Insect Pests Infesting Chilli (*Capsicum annum* L.). *Int. J. Pure App. Biosci.*, 5(6): 1465-1470.

Ravi, K.C., Mohan, K.S., Manjunath, T.M., Head, G., Patil, B.V., Greba, D.P.A., Premalatha, K., Peter, J. and Rao, N.G.V. (2005). Relative Abundance of *Helicoverpa armigera* (Lepidoptera: Noctuidae) on Different Host Crops in India and the Role of These Crops as Natural Refuge for *Bt* Cotton. *Environmental Entomology*, 34(1): 59–69. doi:10.1603/0046-225x-34.1.59

Razak, T.A., Santhakumar, T., Mageswari, K. and Santhi, S. (2014). Studies on efficacy of certain neem products against *Spodoptera litura* (Fab.). *J.Bio.Pest.*, 7(Supp.):160-163.

Reed, W. and Pawar, C.S. (1981). *Heliothis*: A Global Problem. In: Reed W, Kumble V. (eds.) Pro ceedings of the International Workshop on *Heliothis* Management, 15-20 November 1981, Patancheru, India. International Crops Research Institute for the Semi-Arid Tropics; 1982. 9-14.

Renkema, J.M., LeFors, J.A. and Johnson, D.T. (2017). First report of broad mite (Acari: *Tarsonemidae*) on commercial strawberry in Florida. *Florida Entomologist*, 100(4):804.

Revathi, N., Ravikumar, G., Kalaiselvi, M., Gomathi, D. and Uma, C. (2011). Pathogenicity of Three Entomopathogenic Fungi against *Helicoverpa armigera*. *J. Plant Pathol. Microbiol.*, 2:114. doi:10.4172/2157-7471.1000114.

Romero, M.D. and Peña, J.E. (1998). Relationship of broad mite (Acari: Tarsonemidae) to host phenology and injury levels in *Capsicum annuum*. *Florida Entomologist*, 81(4): 515-526.

Saini, A., Ahir, K.C., Rana, B.S. and Kumar, R. (2017). Population dynamics of sucking pests infesting chilli (*Capsicum annum* L.). *Journal of Entomology and Zoology Studies*, 5(2): 250-252.

Sakimura, K. (1932). Life History of *Thrips Tabaci* L. on *Emilia sagittata* and Its Host Plant Range in Hawaii. *Journal of Economic Entomology*, 25(4): 884–891.

Sapkal, S.D., Sonkamble, M.M. and Gaikwad, B.B. (2018). Seasonal incidence of tomato fruit borer, *Helicoverpa armigera* (Hubner) on tomato, *Lycopersicon esculentum* (Mill) under protected cultivation. *Journal of Entomology and Zoology Studies*, 6(4): 1911-1914.

sci-hub.tw/10.2903/j.efsa.2019.5620

Shahout, H.A., Xu, J.X., Jing, Q. and QiaoDong, J. (2011). Sublethal effects of methoxyfenozide, in comparison to chlorfluazuron and beta-cypermethrin, on the reproductive characteristics of common cutworm, *Spodoptera litura* (Fabricius) (Lepidoptera: Noctuidae). Journal of the Entomological Research Society, 13(3): 53-63. http://www.entomol.org

Sinha, S.H. (1993). Neem in the integrated management of *Helicoverpa armigera* Hubn. in chickpea. *World Neem Conference,* Bangalore, India 24-28 Feb. 1993. pp 6.

Story, R.N. and Keaster, A.J. (1982a). The overwintering biology of the black cutworm, *Agrotis ipsilon,* in field cages (Lepidoptera: Noctuidae). *Journal of the Kansas Entomological Society,* 55: 621-624.

Story, R.N. and Keaster, A.J. (1982b). Temporal and spatial distribution of black cutworms in midwest field crops. *Environmental Entomology,* 11: 1019-1022.

Subramanian, S. and Mohankumar, S. (2006). Genetic variability of the bollworm, *Helicoverpa armigera,* occurring on different host plants. *Journal of Insect Science,* 6(26): 1–8. doi:10.1673/2006_06_26.1

Sumit, H.K., Sundar, B., Sandhya, S. and Sharma, A.K. (2019). Study on seasonal incidence of insect pests of vegetable crops collected through light trap. *International Journal of Chemical Studies,* 7(4): 687-689.

Sundar, B., Rashmi, V., Sumith, H.K. and Sandhya, S. (2018). Study the incidence and period of activity of *Spodoptera litura* on soybean. *Journal of Entomology and Zoology Studies,* 6(5): 331-333.

Van Maanen, R., Vila, E., Sabelis, M. W., and Janssen, A. (2010). Biological control of broad mites (*Polyphagotarsonemus latus*) with the generalist predator *Amblyseius swirskii. Experimental and Applied Acarology,* 52(1), 29–34. doi:10.1007/s10493-010-9343-2.

Van Rijn, P., Mollema, C. and Steenhuis-Broers, G. (1995). Comparative life history studies of *Frankliniella occidentalis* and *Thrips tabaci* (Thysanoptera: Thripidae) on cucumber. *Bulletin of Entomological Research,* 85(2): 285-297.

Vashishtha, S., Chandel, Y.S., Chandel, R.S. (2018). Comparative efficacy of indigenous heterorhabditid nematodes from north western Himalaya and *Heterorhabditis indica* (Poinar, Karunakar and David) against the larvae of *Helicoverpa armigera* (Hubner). *International Journal of Pest Management,* 65(1): 16-22.

Vashishtha, S., Chandel, Y.S., Kumar, S. (2012). Biology and damage potential of *Spodoptera litura* Fabricius on some important greenhouse crops. *Journal of Insect Science,* 25(2): 150-154.

Velayutham, L.K. and Damodaran, K. (2015). An economic analysis of chillies production in Guntur district of Andhra Pradesh. *International Journal of Research in Economics and Social Sciences,* 5(9): 43-49.

Walter D.E. and Proctor H.C. (2013). Life Cycles, Development and Size. In: Mites: Ecology, Evolution and Behaviour. Springer, Dordrecht.

War, A.R., Paulraj, M.G. and Ignacimuthu, S. (2011). Synergistic Activity of Endosulfan with Neem Oil Formulation against Tobacco Caterpillar, Spodoptera litura (Fab.) (Lepidoptera: Noctuidae). *Journal of Entomology*, 8: 530-538.

Weber, G. (1985). Genetic variability in host plant adaptation of the green peach aphid, *Myzus persicae*. *Entomologia Experimentalis et Applicata*, 38(1): 49–56. doi:10.1111/j.1570-7458.1985.tb03497.x

Xue, M., Pang, Y.-H., Wang, H.-T., Li, Q.-L. and Liu, T.-X. (2010). Effects of Four Host Plants on Biology and Food Utilization of the Cutworm, *Spodoptera litura*. *Journal of Insect Science*, 10(22): 1–14. doi:10.1673/031.010.2201.

Zalucki, M.P.; Murray, D.A.H.; Gregg, P.C.; Fitt, G.P., Twine, P.H. and Jones, C. (1994). Ecology of *Helicoverpa armigera* (Hübner) and *H. punctigera* (Wallengren) in the inland of Australia: larval sampling and host plant relationships during winter and spring. *Australian Journal of Zoology*, 42: 329-346.

<u>Flower Crops</u>

16

Biointensive Integrated Pest Management of Commercial Flower Crops

Davendra Kumar, Sapna Panwar and Girish K.S.

Introduction

Biointensive integrated pest management (BIPM) is essentially a component of integrated pest management which helps to reduce the dependence on chemical pesticides thereby preventing ecological deterioration. The major elements of this approach includes many factors viz. host plant resistance, use of beneficial organism, agronomic practices, bio pesticides, parasitoids, predators, plant exudates etc. It also includes all other conventional non-chemical methods of pest control like use of botanical insecticides preferably plant based insecticides like Azadirachtin from Neem (*Azadirachta indica*), Rotenone from *Derris chinensis* and *Derris elliptica*, Nicotine from Tobacco (*Nicotiana tabacum*), Juvocimene 1 or 2 and Ocimin from *Ocimum sanctum*, Atropine from *Datura stramonium*, Annonine and Squamocin from *Annona squamosa* etc.

Advantages of Biointensive Integrated Pest Management

1. Reduces the need for pesticides by using several pest management methods
2. Reduces or eliminates issues related to pesticide residue
3. Maintains or increases the cost-effectiveness of a pest management program
4. Promotes sustainable bio-based pest management
5. Protects non-target species through reduced impact of pest management activities

Table 1. Major insect pests of commercial flower crops

Crop	Insect-Pest		Order
Rose	Red scale	*Aonidiella aurantii*	Hemiptera
	Thrips	*Scirothrips dorsalis*	Thysanoptera
	Aphids	*Macrosiphum rosae*	Hemiptera
	Hairy caterpillar	*Euproctiesfracterna*	Lepidoptera
	Bud borer	*Helicoverpa armigera*	Lepidoptera
	Chafer beetle	*Adoretus* spp.	Coleoptera
	Termites	*Microtermes obesi*	Isoptera
Marigold	Bud caterpillar	*Helicoverpaarmigera*	Lepidoptera
	Aphids	*Aphis gossypii*	Hemiptera
	Thrips	*Thrips tabaci*	Thysanoptera
Carnation	Bud borer	*Helicoverpaarmigera*	Lepidoptera
	Thrips	*Haplothripsspp.*	Thysanoptera
	Two spotted spider mites	*Tetranychusurticae*	Trombidiformes
Chrysanthemum	Aphids	*Macrosiphoniellasanborni*	Hemiptera
	Thrips	*Microcephalothripsabdominalis*	Thysanoptera
	Leaf folder	*Hydyleptaindicata*	Lepidoptera
	Leaf miner	*Liriomyzatrifolii*	Diptera
Gladiolus	Thrips	*Taeniothrips simplex*	Thysanoptera
	Cut worms	*Agrotissegetum*	Lepidoptera
	Leaf caterpillar	*Spodopteralitura*	Lepidoptera
	Mites	*Tetranychusspp.*	Trombidiformes
	Mealy bug	*Ferrisiavirgata*	Hemiptera
Tuberose	Bud borer	*Helicoverpaarmigera*	Lepidoptera
	Thrips	*Thrips* spp.	Thysanoptera
	Weevils	*Myllocerusspp.*	Coleoptera
	Pentatomid bug	*Nezaraviridula*	Hemiptera
	Aphids	*Aphis* spp.	Hemiptera
Jasmine	Bud worm	*Hendecasisduplifascialis*	Lepidoptera
	Blossom midge	*Contariniamaculipenis*	Diptera
	Gall mite	*Aceria jasmine*	Trombidiformes
Gerbera	Whitefly	*Bemisiatabaci*	Hemiptera
	Leaf miner	*Liriomyzatrifolii*	Diptera
	Thrips	*Thrips palmi*	Thysanoptera
	Aphids	*Myzuspersicae*	Hemiptera
	Yellow mite	*Polyphagotarsonemus latus*	Acari

Mealy bugs on chrysantheum

Mealy bugs on chrysantheum

Mealy bugs on tuberose

Aphids on lily

Aphids on chrysantheum

Aphids on gladiolus

White fly on jasmine

Aphids on rose

Scale insects on rose

Fig. 1. Damage symptom of insect pests on commercial flower crops

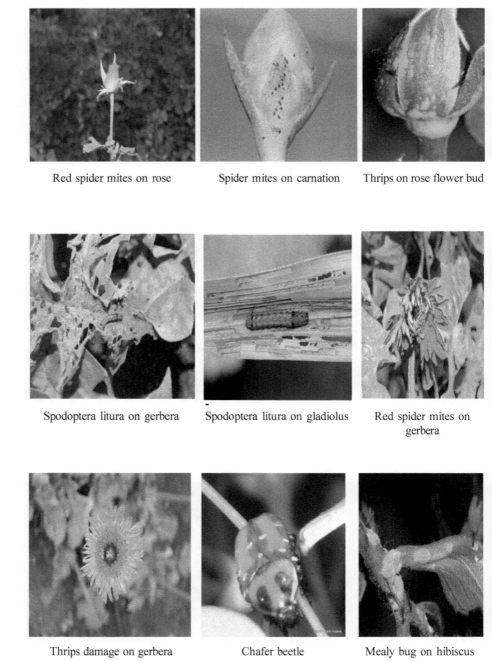

Fig. 2. Damage symptom of insect pests on commercial flower crops

Bud borer on Rose	Bud borer on China Aster
Mealy bug on rose	Hairy caterpillar on Chrysanthemum

Fig. 3. Damage symptom of insect pests on commercial flower crops

General Biointensive Integrated Pest Management Strategies on Commercial crops

The most recent Biointensive integrated strategy for management of pest includes several components like physical methods which include hot water treatment of planting material, soil solarization and bio-rational chemicals like pheromones. The cultural methods includes crop rotation, summer ploughing, fallowing the land, intercropping, pruning, mulching, spacing, planting date, trap cropping, use of resistant cultivars, etc. Biological methods includes use of bio-control agents (predators, parasitoids and mycorrhizal fungi), botanicals based including bio-fumigation, oil cakes, FYM, crop residues, green manuring and other organic amendments.

A. Cultural methods

Uses of different cultural has been advocated to check pest populations by altering the planting time, increasing the plant diversity, use of trap, barrier and inter crops, field sanitation, fertilizer as well as water management and crop rotation etc., have been advocated. For example early sown crop marigold suffers lower incidence of *Helicoverpa armigera* (Srinivasan and Krishnamoorthy, 1993).

B. Physical methods

Pheromone is a substance that is released into the environment by a member of a species that produces a specific response in members of the same species. For example sex pheromone (Septa) are insect specific produced artificially by the females in a laboratories which attracts the males of the same species over a longer distance and aggregation pheromone which is produced by male that attract both males, females, adults and larvae of same species over a shorter distances.

C. Biological methods

Biological control is the beneficial action of parasites, pathogens, and predators in managing pests and their damage.

Predators are mainly free-living species that directly consume a large number of preys during their whole lifetime. For example, *Chrysoperla zastrowiarabica* is an effective predator for control of white fly, aphids, jassids and eggs for some lepidopterous borers.

Predatory Ladybird beetle

Predatory Syrphid fly Larva   Predatory Spider

Fig. 4. Some predators on commercial crop

D) Botanical methods

Allelochemicals are the secondary metabolites produced by organisms such as plants or microorganisms which are not needed for primary metabolism and this chemicals released from donor plants into the environment and affect growth and development of receiver pests. For example Marigold plant (*Tagetes* spp.) secret alpha-terthienyl from the root of *Tagetes* spp. that repels the several nematodes such as root-knot nematodes (*Meloidogyne* spp.) and *Pratylenchus* spp. and *Chrysanthemum cinerarifolium* produce a chemical called Pyrethrum which is used for reduction of whitefly, aphid and hornworm etc and where as Botanical insecticides are naturally occurring chemicals (insect toxins) extracted or derived from plants that have naturally occurring defensive properties.

Management of major pest on flower crops using Biointensive IPM

1. *Helicoverpa armigera*

Damage symptom

Larvae prefer to feed on reproductive parts of plant (flowers and buds). It may also feed on foliage. Due to Feeding results in holes bored into reproductive structures and feeding within the plant. It may be necessary to cut open the plant organs to detect the pest. Secondary pathogens like fungi, bacteria may develop due to the wounding of the plant. Frass may occur near to feeding hole from larval feeding within.

a) Mechanical methods

- Removal and destruction of crop residues after harvest to avoid the carry over population of *Helicoverpa armigera* to next season.
- Hand picking and destruction of various insect stages, affected plant parts and rosetted flowers.
- Removal of terminal shoot of marigold (Pinching) at 30-40 days after planting to reduce *Helicoverpa* oviposition and also to encourage sympodial branching which help to bears more number of flower.
- Removal and destruction of alternate weed hosts of *Helicoverpa armigera* like *Abutilon indicum, Chrozophore rottlari and Solanum nigrum.*

b) Cultural methods

- Summer deep ploughing to expose soil inhabiting / resting stages of insects and nematode population.
- Use of trap crops like okra, marigold (*Tagetes* spp.), early pigeon pea, coriander, maize crops along the borders is recommended. Insects feeding on these crops must be removed and destroyed.

- Adopt proper crop rotation
- Adopt proper spacing, irrigation and fertilizer management. Avoid application of excess nitrogenous fertilizers.

c) Botanical methods

- Spray neem seed kernel extract (NSKE) 5% as a strong oviposition deterrent or neem oil 3% starting from one month after planting at 15 days interval.
- If the infestation is severe, spray the botanicals insecticide such as Azadirectin @ 1500 ppm at weekly intervals.

d) Biological methods

- Application of *Helicoverpa armigera* nuclearpolyhedrosis virus (HaNPV) @ 250-500 ml/ha depending upon the crop growth with jaggery and teepol in evening hours at 7th and 12th week after sowing.
- ULV spray of NPV at 3×10^{12} POB /ha with 10% cotton seed kernel extract (CSKE), 10% crude sugar, 0.1% each of Tinopal and Teepol.
- Inundative release of egg parasitoid, *Trichogramma* spp., at 6.25 cc/ha at 15 days interval 3 times from 45 DAS.
- Inundative release of egg-larval parasitoid, *Chelonus blackburnii* and predator, *Chrysoperla carnea* at 100000/ha at 6th, 13th and 14th weeks after sowing.
- *Trichogramma brassiliensis*, 2.5 cc/ha. once in 10 days (Egg parasitoid).

e) Behavioural methods

Set up pheromone traps @ 12/ha for monitoring *Helicoverpa armigera* at a distance of 50 m. When the trapped moths are 8/day necessary action may be taken. Trapped moths should be removed daily.

2. Aphids

Damage symptom

- Aphid suck cell sap from tender leaves, buds and twigs
- The feeding results in disfiguring, leaves curling downward and withering flower
- A black fungus may also develop on the honey dew secreted by aphids.

a) Mechanical methods

- Collection and destruction of aphid infested twigs.

- Remove the little leaf affected plants.
- Clipping the tips of the seedlings up to 2 inches prior to flowering to remove the egg masses of aphids if any.

b) Cultural methods

- Growing castor along the borders and irrigation bunds as trap crop for aphids.
- Deep summer ploughing on bright sunny days during the months of May or June should be done to expose soil inhabiting or resting stages of insects and nematode population. The field should be kept exposed to sunlight for at least 2-3 weeks.

c)   Botanical methods

- Spray fish oil resin soap 25 kg / ha @1 kg in 40 litre of water.
- Spray neem oil 3% plus teepol (1 ml/litre) or spray neem seed kernel extract 5 %.
- Spray NSKE (5%) or neem oil (3%) alternatively.
- Apply neem cake @ 250 kg/ha as a basal dosage.

d) Biological methods

- *Chrysoperla* spp. (Green lace wing) 5000-10000 eggs /ha after starting of first sign of aphid presence 3–4 times in 15 days.
- Application of *Verticillium lecanii* (0.5-1.0%) affects the all growth stages of aphid.
- Conserve the predator's *viz.,* spiders, coccinellids and wasps to check the population of aphids.

e)   Behavioural methods

- Use of light trap @ 12/ha to monitor and trap the aphids.
- Set up the yellow sticky traps @ 25/ha to monitor the activity of pest and to synchronise the botanical pesticide application, if need be, at the maximum activity stage.

3. White fly

Damage symptom

Adult white flies resemble tiny white moths. They are about 1/16th of an inch long, white, with four wings, and may be covered with a powdery white wax.

The nymphs or immature whiteflies resemble flat disks on the undersides of the leaves and may or may not be covered with wax. White flies injure plants by sucking plant sap. They excrete honeydew, which gives the plants a sticky appearance and supports the growth of black sooty mold. Heavy white fly infestations can cause stunting, distorted, discolored leaves and a plant that appears unthrifty.

a) Mechanical methods

- Removal and destruction of alternate weed hosts of white fly like *Abutilon indicum, Chrozophore rottlari, Solanum nigrum* and *Hibiscus ficulensus* from the fields and neighbouring areas and maintaining field sanitation.

- Use aluminum foil or reflective plastic or silver coloured mulches that can repel adult whiteflies.

b) Cultural methods

- Collection and destruction of leaves infested with white fly.

- Keep target areas free of weeds that can serve as whitefly hosts.

- Avoid excess use of nitrogen fertilizer, including manures as succulent growth will increase whitefly population.

c) Botanical methods

- Spray NSKE 5% or neem oil (5 ml/litre) or 5 % notchi leaf extract or 5% *Catharanthus rosea* extract.

- Spray fish oil rosin soap 25 g/ litre and add teepol as wetting agent

- If the infestation is severe, spray the botanicals insecticide such as Azadirectin @ 1500 ppm at weekly intervals.

d) Biological methods

- *Chrysoperla* spp. (Green lace wing) 5000-10000 eggs /ha, 3-4 times in 15 days after first sign of whitefly presence.

- *Delphastus pusillus* is the most whitefly-specific predator. Recommended release rates are 7-10 per m².

- *Eretmoceruseremicus* is the most effective parasitoid. Recommended release rates in the greenhouse are three wasps per m² every 1-2 weeks after starting of first sign of whitefly presence.

e) Behavioural methods

- Monitoring the activities of the adult white flies by setting up yellow pan traps and sticky traps @ 12/ha at 1 foot height above the plant canopy.

- Locally available empty yellow palmoline tins coated with grease / Vaseline/ castor oil on outer surface may also be used to catch white flies.

4. Thrips

Damage symptom

Thrips feed by piercing plant cells with their mouth parts and sucking out their contents and always feeds on tender part of the plant, such as buds, flowers, or leaves. The effect of their numerous but shallow punctures is to give the injured tissue a shrunken appearance. They feed on the thick fleshy petals, pistils, and stamens of the flower, and then the affected parts turn brownish-yellow, blacken, shrivel up, and drop prematurely. Infested rose blossoms turn brown, and buds open only partially. The petals, distorted with brown edges, seem to stick together. Only the epidermis and relatively few mesophyll cells are affected.

a) Mechanical methods

- Collection and destruction of leaves infested with thrips
- Removal and destruction of nearby weed which acts as alternate hosts of thrips from the fields, neighbouring areas and maintaining field sanitation.
- Use silver or grey coloured mulches of 100 micron or 400 gauges that can repel thrips.

b) Cultural methods

- Keep plants well irrigated and avoid excessive applications of nitrogen fertilizer which may promote higher populations of thrips
- Old and spent flowers can harbour thrips, so their removal and disposal is sometimes recommended.
- Grow resistant cultivars. For example, thrips more often damages fragrant, light-coloured, or white roses so we need to recommended pink, red or yellow colour roses.
- Avoid close planting in thrips prone areas or during summer seasons.
- Prune and destroy injured and infested terminals shoots.

c) Botanical methods

Use neem extract @1500 ppm, if pest occurrence is very severe.

d) Biological methods

- *Chrysoperla* spp. (Green lace wing) 5000-10000 eggs /ha after starting of first sign of thrips presence 3–4 times in 15 days.
- Usepredatory mites such as phytoseiid mites (*Amblyseius* spp.) and pirate bugs (*Orius* spp.).

e) Behavioural methods

Set up the blue sticky traps @ 20-25/ha to monitor the activity of pest and to synchronise the botanical pesticide application, if need be, at the maximum activity stage.

5. Mites

Damage symptom

Two spotted spider mite and broad mites are important mites that can cause problems on flowers. These mites are most active on the underside of the leaves, their presence being apparent by the fine stippling caused by their feeding and seen on the upper surface of the leaves and leave curl downward and turn coppery or purplish. Fine webbing is produced by this mites and leaves turn yellow or bronze, and many drop.While broad mites are invisible to the naked eye but cause a great deal of damage, particularly to Dahlia, Chrysanthemum and Verbena.Mite feeding causes the leaves to curl, twist, and become brittle and scabby. Flower buds may dry up and die. Light infestations may result in discolored or dark-flecked flowers.

a) Mechanical methods

- Flooding the growing area to provide the proper moisture to soil surface.
- Removal and destruction of crop residues after harvest to avoid the carry over population of mites to next season.

b) Cultural methods

- Growing areas should not be dry provide adequate water especially during of drought period, spider mites will often be worse on drought stressed plants first.
- Inside greenhouses maintain higher humidity levels that can reduce spider mite populations and damage.

c) Botanical methods

Spray neem seed kernel extract (NSKE) @5% or neem oil @3%.

d) Biological methods

- The number one predator for two spotted spider mites is the persimilis mite (*Phytoseiulus persimilis*). This mite works best at temperatures between 55°F to 85°F. The recommended release rate is 1 mite per square foot for effective controlling.
- During hot summer months, the swirski mite (*Amblyseius swirskii*) works best on two spotted spider mites. Swirskii works best at temperatures above

68°F. They are very aggressive and can be applied generally at a rate of 5-10 mites per square foot.

- Application of *Beauveria bassiana* (1.0%) affects the young stage of mites.
- Application of *Bacillus thuringiensis* subsp. *Kustaki* (0.3-0.4%) affects all stage of mites.

e) Behavioural methods

- Set up *Bacillus thuringiensis* mite trap @ 12/ha for monitoring mite at a distance of 40-60 m is beneficial.

References

Alavo, T.B. (2006). Biological control agents and environmentally-friendly compounds for the integrated management of *Helicoverpa armigera* Hübner (Lepidoptera: Noctuidae) on cotton: Perspectives for pyrethroid resistance management in West Africa. *Archives of Phytopathology and Plant Protection*, 39(02): 105-111.

Butter, N.S., Singh, G. and Dhawan, A.K. (2003). Laboratory evaluation of the insect growth regulator lufenuron against *Helicoverpaarmigera* on cotton. *Phytoparasitica*, 31(2): 200-203.

Campos, E.V., Proença, P.L., Oliveira, J.L., Bakshi, M., Abhilash, P.C. and Fraceto, L.F. (2019). Use of botanical insecticides for sustainable agriculture: future perspectives. *Ecological Indicators*, 105: 483-495.

Chadha, K.L. (2019). *Handbook of Horticulture in Vol 2*. New Delhi: Indian Council Agricultural Research.

El-Wakeil, N.E. (2013).Retracted article: Botanical Pesticides and Their Mode of Action. *GesundePflanzen*, 65(4): 125-149.

Mohan, S., Devasenapathy, P., Vennila, S. and Gill, M. S. (2015).Pest and disease management: organic ecosystem. Coimbatore: Tamil Nadu Agricultural University.

Mouden, S., Sarmiento, K.F., Klinkhamer, P.G., and Leiss, K.A. (2017). Integrated pest management in western flower thrips: past, present and future. *Pest Management Science*, 73(5): 813-822.

Nivsarkar, M., Cherian, B. and Padh, H. (2001). Alpha-terthienyl-A plant derived new generation insecticide. *Current Science, Bangalore*, 81(6): 667-672.

Srinivasan, K., Krishnamoorthy, P.N. and Prasad, R. (1993). Evaluation of different trap crops for the management of fruit borer, H. *armigera* on tomato. In Abstract, Golden Jubilee Symposium. Horticultural Research: Changing Scenario Bangalore, India. 259pp.

Medicinal, Aromatic and Spice Crops

17

Insect-Pest of Common Medicinal and Aromatic Plants and Their Sustainable Management

Umesh Das, Suprakash Pal and Nagendra Kumar

Introduction

Medicinal plants are known to Indian traditional healers since time immemorial. The plants were collected from the wild and were used in many preparations of wellness products. India is endowed with diverse group of medicinal plants accounting for more than 8000 species which are being used in more than 10,000 herbal products. Ninety percent of herbal industry's requirement of raw materials is meted out from the natural ecosystem-forests-resulting in ruthless exploitation and destruction of its natural habitats (Mathivanan *et al.*, 2016). Medicinal plants are cultivated or found in the wild throughout the country and are used for various purposes including traditional herbal remedy (infusion, tincture and decoction) and extraction of phytochemicals for homoeopathic and ayurvedic drugs, cosmetics, neutraceuticals/dietary supplements, functional foods and aroma therapy oils (Nagpal *et al.*, 2004). The large scale cultivation of medicinal plants may face the problem of sudden appearance of large populations of insect pests in a single crop. Like other plants, medicinal plants too have to bear the devastating effects of injurious insect-pests, which are not only harmful for the plant but also, deteriorate the quality of the produce, thus hampering its medicinal value.This chapter discusses major insect-pest and their eco-friendly measures in medicinal plants so that a package of practices can be recommended for an easy adoption even by small and marginal farmers and gardeners in India.

Pests of common medicinal plant

A. Tulsi, (*O. sanctum*)

Tulsi crop is attacked by several sucking insect pests like lace bug *Cochlochila bullita* (Stal), whitefly *Aleurodicus dispersus* Russell, *Dialeurodes* sp. and aphid *Macrosiphum* sp. but *C. bullita* and *A. despersus* found throughout the year while aphid appeared occasionally in December-January (Sathe *et al.*, 2014). Among these pests, *C. bullita* caused severe damage. Besides these pests, basil plant is also attacked by other pestsinsects like leaf roller, lace bug, etc. (Panda, 2005).

1. Lacewing Bug, *Cochlochila bullita* (stal) (Hemiptera: *Tingidae*)

Lacewing bug is one of the serious pests of Tulsi, adult lace bugs usually feed on tender shoots of the herb causing them to wilt and eventually die and in many instances, nymphs and adults feed, gregariously on the leaves, leaving tiny black spots of excrement on the upper surface of the leaves (Dhiman and Jain, 2010). *O. basilicum* plant is severely infested by the population of *C. bullita*at Saharanpur and causes drying and wilting of leaves resulting in ultimate death plants (Dhiman and Datta, 2013). According to Sajap and Peng (2010), adult *C. bullita* usually feed on tender shoots of the herbs causing them to wilt and eventually die. Nymphs are found on the under surface while the adults are on upper surface in colonies on the tender foliage and shoots. Leaves become discoloured and gradually dry up. Nymphs were observed on upper and lower leaf surfaces and reproductive structures, with foliar chlorosis evident on the upper surfaces and dark excrement on lower surfaces, severe feeding resulted in the death of plants. *C. bullita* was responsible for sucking the cell sap and there by curling the leaves of *Orthosiphon stamineus* (Sajap and Peng, 2010).

Butani (1982) suggested, *O. kilim* and-*scharicum* and Tigvattnanont (1989) *O. bacilicum* and *Co-leus parviflorus* as host plants for *C. bullita*. Arvind Kumar (2013) reported *C. bullita*on *O. sanctum* from Zharkhand. However, no pests have been reported as cell sap suckers or defoliators on *O. sanctum* from Kolhapur region, Western Maharashtra.

Fig. Lacewing Bug

2. White fly, *A. despursus, Di aleurodes* sp. (Hemiptera : *Aleurodidae*)

White fly complex which suck the cell sap of tulsi and kill plants, have been reported for the first time from this region. *A. despursus* has never been reported on *O. sanctum* from India. However, Sahte & Margaj, (2001) reported *A. despersus* on cotton, to mato, mulberry, guava, china rose and brinjal from Kolhapur region. Sathe *et al.*, (2014) reported that *A. dispursus* and *Dialeurodes* completed their life cycles from egg to adult within 22.5 days and 190 days respec-tively on *O. sanctum*. *A. despursus* incidence on tulsi was also throughout the year. However, the *Dialerodes* sp. was found on the crop only from June to November and then from March to May with moderate or low populations.

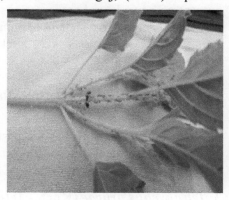

Fig. Aphid

3. Aphid, *Macrosiphum* sp. (Hemiptera : *Aphididae*)

Aphid are minute green in colored, infesting the foliage and tender shoot of Tulsi. The nymphs and adults desap the plant and secreted honey dew like substance, created sooty moulds and affected photosynthesis, growth and yield of the crop.

Management

Sucking pests can be controlled by spraying the crop with 0.03 Azadiractin or Neem oil 0.5.

B. Ashwagandha (*Withania somnifera*)

Ramanna (2009) documented 11 species of pests on ashwagandha which included 5 species of defoliators (Coleoptera and Lepidoptera), 5 species of sucking pests (Hemiptera, Acarina) and 1 species of flower and fruit feeder (Lepidoptera). Amongthe different species of pests observed, the defoliators, viz. *Henosepilachna vigintioctopunctata* Fab., *Deilephila nerii* Linn., *Myllocerus viridanus* Fab., *Myllocerus discolor* Fab. and leaf miners, were found occasionally at initial stagesof the crop growth causing reduction in foliage. The sucking pests, viz. *Aphis gossypii* Glover, mealybug, *Ferrisiavirgata* Fab., green plant bug and *Nezara viridula*, were observed during different growth periods. The lepidopteran caterpillar, *Helicoverpa* sp., was found feeding on leaves and also boring into flowers andfruits. Mite, *Tetranychus urticae* Koch, caused yellowing and drying of leaves which ultimately reduced the yield.

1. Tree hoppers, *Leptocentrus substitutes* (Hemiptera : *Membracidae*)

Both nymph and adults of the tree hopper caused damage to the ashwagandha by sucking the sap from the leaves and shoots which resulted in yellowing and devitalisation. These hoppers are small, had a backwardly directed spine like process with two lateral processes on pronotum (Rehaman *et al.*, 2018). These results are similar to the findings of Suman and Swaminathan (2007) who reported *Leptocentrus substitutes* were sucking the sap from the tender parts of ashwagandha plants.

2. Green plant bug, *Acrosternum gramineum* (Fab.) (Hemiptera: *Pentatomidae*)

Green plant bug is polyphagous pest, where both nymphs and adults were observed to suck the sap from leaves, buds and young shoots of ashwagandha causing withering of berries.

3. *Tetranychus urticae* (Koch.) Tetranychiidae

A mite, *Tetranychus urticae* (Koch.) caused damage to the leaves of ashwagandha (Rehaman *et al.*, 2018). Severe infestation leads to yellowing and stunting of the plant. These finding are lined with Indranil *et al.*, (2008) and recorded phytophagous mite *T. urticae* on ashwagandha from West Bengal

Fig. *Tetranychus urticae*

4. Fruit borer, *Helicoverpa armigera* (Hub.) (Lepidoptera:*Noctuidae*)

Fruit borer, *Helicoverpa armigera* (Hub.) is apolyphagous pest. Larvae of *Helicoverpa armigera* enters the fruit and feed while remaining concealed inside the fruit. They bore into the developing fruits causing severe damage and thereby reducing the marketable yield (Rehaman *et al.*, 2018).

5. Sphingid, *Deilephila nerii* (Linn.) (Lepidoptera: *Sphingidae*)

The larvae of Sphingid, *Deilephila nerii* (Linn.) were found defoliating the leaves of ashwagandha with population of 0.58 larvae per plant (Rehaman *et al.*, 2018).

6. Spotted beetle, *Henosepilachna vigintioctopunctata* (F.) (Coleoptera: *Coccinellidae*)

The grubs and adults scrap out and feed on the green tissues of the leaves which ultimately dries up. Chaudhary and Saravanan (2011) recorded Spotted beetle as one of the major pest of ashwagandha in Gujarat. Its seasonal incidence incultivated ashwagandha during 2008-09 and 2009-10 was studied. Spotted beetle wasactive from last week of September to mid of December with a peak population (7.8 grubs/plant) and (1.73 grubs/plant) in 2nd week of October (41 SMW) during both the years of observation.

Management

- Application of farmyard manure (FYM) (12.5 t/ha) + Azophos (2 kg/ha) + neem cake (1000 kg/ha) was found to be very effective in reducing the damage of spotted leaf beetle by 69.79 per cent.

- FYM + Azophos + neem cake combination was less preferred for oviposition which recorded 62.00 eggs/plants, coupled with a minimum feeding area (6.75 cm2) (Ravikumar *et al.*, 2008).

- NSKE (5%) showed the lowest infestation by the epilachna beetle and highest yield (Chandranath and Katti, 2010).

- *B. thuringensis* was found to be the most effective against red bug and aphid. *Chrysoperla cornea* @ 500 per 100 m² followed by parasitoid *Trichogramma chilonis* at 1500 per 100 m²) was also found to be the most effective for reduction of the larval population of defoliators (Meshram *et al.*, 2015).

- *A. gossypii* and increased crop yield by 40 per cent with cost-benefit (CB) ratio of 1:16.8. The highest mean per cent reduction of H. armigera infestation on ashwagandha was recorded with Ha NPV at 250 LE (63.61), which was on par with nimbicidine at 3 ml/L (56.66) followed by NSKE 5 per cent (47.42) and neem oil 5 ml/l (45.73) over control (Rehaman and Pradeep, 2016).

- The gram caterpillars, *H. armigera* bore into fruits and devour tender leaves of ashwagandha plants. As an integrated measure field release of *T. chilonis* (100,000 eggs/ha), spraying of *B. thuringiensis* (0.5 kg/ha) or HaNPV (250 LE/ha) and installation of pheromone traps (12 traps/ha) was suggested. Further, the hawk moth caterpillar, *Deilephila nerri* a minor defoliator, is controlled by soil digging to expose pupae, hand picking and destruction of larvae, installation of light traps (1 trap/ha) and planting of nerium (*Nerium oleander* L.) as trap plant around the field (TNAU, Agri portal, 2016).

C. Sarpagandha, *Rauvolfia serpentina* (Benth.)

Rehaman *et al.* (2018) reported that several insect pest attacking the leaves and berries of Sarpagandha and causes economic damage such as lablab bug, grasshopper, curculionid weevil and sphingid larvae. Gogoi *et al.* (2017) reported

Rauwolfia serpentine is important medicinal plant and infested by number of insect pest like scale insect and mealy bugs were found abundantly.

1. Scale insect (*Saissetia sp.*) Hemiptera : Coccoidea

Scale insect (*Saissetia sp.*) was found to be the most devastating pest in the western districts of Assam. Hemispherical scales were found clustering on the shoots, leaves, and young fruit of sarpagandha plants. It infects lower surface of leaves along with the mid ribs and mid veins. They are also found on petioles and tender shoots. Scale feeds on the plant sap

Fig. Scale insect

and causes loss of vigour. Leaves turn yellow dries up and drop down resulting asick plants. These toxins in the saliva of scale insect also induced brown spots on the foliage (Gogoi *et al.*, 2017).

2. Mealy bug *Corcidohystrix sp.* (Hemiptera : *Pseudococcidae*)

Another important insect that has been recorded in sarpagandha was mealy bug *Corcidohystrix sp.* The nymphs and adults of mealy bug remain on the under surface of the leaves and stems and suck the sap resulting in deformation of the leaves and branches. Affected plants turned yellow, followed by wilting and drying. Leaves showed crinkling appearance. Bugs secrets honey dew that favors development of shootymould which in turn affects the photosynthesis process by the leaves. Presence of honey dew also invites ants (Gogoi *et al.*, 2017).

Fig. Mealy bug

3. Lab lab bug, *Riptortus pedestris* (Fabricus) (Hemiptera: *Alydidae*)

Lab lab bug, *Riptortus pedestris* (Fabricus) is a dark brown bug, hind femur provided with row of spines. Adults were observed sucking the juice from the berries of *R. serpentaina*. The berries showed withering and drying up symptoms due to the bug attack with mean population of 1.1 bugs per plant (Rehaman *et al.*, 2018).

4. Grasshopper *T. annulata* (Orthoptera: *Acrididae*)

The nymphs and adults of grasshopper were found defoliating the leaves of sarpagandha. It is brown to dark grey with black spots, tegmina with two to three darker bands (Rehaman *et al.*, 2018). This was earlier recorded by Reddy *et al.* (1981) who found damage was more severduring vegetative stage at Dharward.

5. Sphingid, *Deilephila nerii* (Linn.) (Lepidoptera: *Sphingidae*)

The larvae of Sphingid, *Deilephila nerii* (Linn.) were greenish in colour which was found defoliating the leaves of Sarpagandha with population of 0.60 larvae per plant (Rehaman *et al.*, 2018).

Management

- Deep ploughing in summer to expose larvae and pupae of black cutworm, *Agrotis ipsilon* to avian predators, 1-2 light traps/ ha or 20 pheromone traps/ha to trap moths, irrigation to expose larvae to predators.

- Drenching collar region of plants in evening hours with NSKE 5% solution could kill the larvae.

D. Lemon grass, *Cymbopogon citratus*

1. Termite, *Microtermes obesi* (Isoptera : *Termitidae*)

These termites are cream colored with dark head. Workers more active in the morning and evening. It is one of major pest of wheat. They damage wheat just after sowing and close to the maturity. Heavily infected plants may dry of clumps or wilt and can pulled out easily.

2. Mealy bug, *Dysmicoccus brevipes* / *Pseudococcus bromeliae* (Hemiptera: *Pseudococcidae*)

Adults and nymphs seen colonizing near base of the clumps during dry months, sucking of juice cause drying of clumps.

3. Stem borer, *Chilotrea* sp. (Lepidoptera: *Pyralidae*)

Caterpillar of *Chilotrea* sp. cause damage to lemon grass and produce death of tillers.

4. Spittle bug, *Clovia bipunctata* Walk. (Hemiptera: *Cercopidae*)

Both adults and nymphs are suck cell sap and causing stunted growth.

Management

- For manage of Mealy bug sanitary measures are to be adopted like the plot should be kept weed free, Remove and destroy all dried up clumps and Localised application of neemoil garlic emulsion (2%) in affected areas.

- Clean cultivation, Deep ploughing, Crop rotation, Trap crops, Inter cropping are some of the components of cultural control

- Termite can manage by frequent irrigation.

- Conservation of lady bird beetle, aphid loins and parasitic fungi are beneficial in control of aphids.

- Spraying SLNPV @ 250 LE/ha is effective for the control of *Spodoptera litura.*

- Mealybugs can be managed by releasing parasitoids *Cryptolaemus* sp.

- Foliar application of NSKE 5% or neem oil 3% at 15 days interval to reduced the infestation

- Spraying pongima oil 3% twice at 15 days interval provides effective control of pest infesting of pest.

Conclusion

A pest-management program could be enhanced by adopting prevention and suppression techniques. The methods that are most effective and environmental friendly have to be selected while implementing Biointensive Integrated Pest Management program in order to minimize the health hazards. Among the pesticides, neem based pesticides prove to be better in reducing the pest population.

References

Aromatic and Medicinal Plants Research Station (KAU), Odakkali, pp 10-14.

Butani, D.K. (1982). Insect pests of tulsi (*Ocimum sanctum* Linnaeus) and their control. *Pesticides*, 16: 11-12.

Chandranath HT. and Katti P. Management of epilachna beetle on ashwagandha. *Karnataka J Agric. Sci.* 2010; 23(1): 171.

Chaudhary, V. and Saravanan, L. (2011). Biology and seasonal incidence of *Hensopilechna vigintioctopunctata* (Coleoptera:Coccinellidae) on Ashwagandha in charotur region of Gujarat. *Pest management in HorticulturalEcosystem.,* 17(2): 132-139.

Dhiman, S.C. and Datta, O. (2013). Seasonal occurrence of *Cochlochila bullita*: Aserious pest of *Ocimum basilicum. Ann. Pl. Protec. Sci.* 21(1): 176-223.

Dhiman, S.C. and Jain, S. (2010). Seasonal occurrence and damage of *Eusarcocoris capitatus,* a pest of *Ocimum sanctum. Ann. Plant Protec. Sci.* 18: 498-499.

Gogoi, B.B., Sarodee B., Rahul, S., Saikia, D.K. and Senapoty, D. (2017). Report of insect pests and diseases in Pippali and Sarpagandha from Assam. *International Journal of Scientific Research,* 6(3): 364-366.

http//www. http://agritech.tnau.ac.in/ TNAU, Agritech portal, 2016, Tamil Nadu Agricultural University, Coimbatore, Tamil Nadu, India.

Indranil, R., Salik, K.G., Goutham, K.S. (2008). Notes on theoccurrence of mites in festing medicinal plants of Darjeeling Himalayas.*Insect Environment.* 14(3): 130-132.

Jat RS, Reddy RN, Bansal R, Manivel P. Good agricultural practices: ashwagandha, isabgol. Extension Bulletin, Directorate of Medicinal and Aromatic Plants Research, Boriavi, Anand, Gujarat, India, 2015.

Kumar, A. (2013). The lace bug *Cochlochila bullita* (Stal.), a destructive pest of *Ocimum sanctum* in Jharkhand, India.Phytoparasitica, DOI10.1007/S 12600-013-0359-0.

Mathivanan, N., Sithanantham, S., Marimuthu, T., Peter, K.V., Rethinam, P., Brahma, S., Peter, P.I., Kirti, S. (2016). Therapeutic and commercial potential of medicinal plants with special focus on *Morindacitrifolia* L. (Noni). *Souvenir cum Abstracts, Second World Noni Congress*, March, SRM University, Kattankulathur, Tamil Nadu.

Meshram PB, Mawai S, Malviya R. Biological control of insect pests of medicinal plants: *Abelmoschus moschatus, Gloriosa superba* and *Withania somnifera* in forest nursery and plantation in Madhya Pradesh, India. *American J Agric.* 2015; 3(2): 47-53.

Nagpal, A. Karki, M. (2004). A study on marketing opportunities for medicinal, aromatic and dyeplants in south Asia. Medicinal and Aromatic Plants Program in Asia, New Delhi, India.

Ramanna, D. (2009). Investigations on pest complex of medicinal plants and their management withspecial reference to ashwagandha, *With aniasomnifera* (Linn.). *M.Sc. (Ag) Thesis*, University of Agricultural Sciences, Dharwad, 88.

Ravikumar R, Rajedran C, Chinniah S, Irulandi RJ, Pandi. Evaluation of certain organic nutrient sources against mealy bug, *Coccidohystrix insolitus* (Green) and the spotted leaf beetle, *Epilachna vigintioctopunctata* Fab. On ashwagandha, *Withamia somnifera*, J Biopest. 2008; 1: 28-31. 29.

Rehaman SK, Pradeep S. Effect of biopesticides on management of *Helicoverpaarmigera* in ashwagandha (*Withaniasomnifera* Linn.). *Pest Manag. Hort. Ecosyst.* 2016; 22(1): 99-102.

Rehaman, S.K., Pradeep, S., Dhanapal, R. and Chandrashekara, G.V. (2018). Survey studies on insect-pests associated with important medicinal plants in

Shivamogga, Karnataka. *Journal of Entomology and Zoology Studies*, 6(1): 848-857.

Sajap, A.S. and Peng, T.L. (2010). The lace bug *Cochlochilabullita*(Stal) (Heteroptera:*Tingidae*), a potential pest of *Orthosiphon stamineus,* Bentham (Lamiales: *Lamiaceae*) in Malaysia. *Insecta Mundi*, 136: 1-5.

Sathe, T.V., Sathe, N.T., Ghodake, D. and Sathe, A. (2014). Sucking insect pests andmedicinal value of Tulsi (*Ocimum sanctum* L.). *Indian J. Appl. Res.* 4(3): ISSN-2249-555X.

Sharma M, Sharma A, Ram N. Bio-efficacy of nonconventional pesticides against *Tetranychus urticae* Koch infesting ashwagandha (*Withania somnifera* Duval). *Pest Manage. Hort. Ecosyst.*, 2013; 19(2): 254-255.

Suman, M., Swaminathan, R. (2007). Bio-ecology andmanagement of *Henosepilachna vigintioctopunctata* (Fab.) (Coleoptera:Coccinellidae) infesting Ashwagandha. *J. Medicinal and Aromatic Pl. Sci.* 29: 16-19.

Tigvattnanont, S. (1989). Studies on bionomics and local distribution of some lace wings in Thailand: 1 Monanthiagloblulifera (Heteroptera :Tingidae). *Kaenkasetkhonkaen Agriculture Journal*, 18: 200-212.

18

Biointensive Integrated Pest Management of Spice Crops

B.S. Gotyal, S. Satpathy and V. Ramesh Babu

Introduction

Biotic stresses, particularly insect pests adversely affect the economical yield potential and the quality of the widely cultivated spice crops. These pests attack the crops at various developmental stages of the crops. For instance, Pollu beetle alone in black pepper is the most destructive pest and causing 30 to 40% yield loss in humid, tropical evergreen forests of India (Devasahayam, *et al.*, 1988). Considering the cost of cultivation and profitability, it is very important to develop and recommend low cost, economic easily adoptable technologies. Food production process in India during the green revolution period has been based on the use of more chemical fertilizers and pesticides which are detrimental to the environment. The challenge before the crop protection scientists is to boost the yield from the available land without harming the environment. To manage the problem of insect pests in spice crops, farmers are over using synthetic chemical pesticides indiscriminately and this result in environmental degradation and high pesticide residue levels in the produces are are mainly hindering the export. Now the focus is on organic spice production, therefore a search for safer measures of pest and disease management is gaining importance. Integration of various approaches like use of resistant/tolerant varieties, disease or pest free planting materials and exploitation of biological means such as bio-control agents, bio-pesticides, entomo-pathogens, parasitoids and predators at the right time paved the way to achieve this goal (Dhanya *et al.*, 2019; Prakash *et al.*, 2016). Incorporating ecological and economic factors into decision making and addressing the public concerns about environmental quality and food safety is the need of the hour. There is now wide array of techniques available to replace the use of conventional chemical insecticides in IPM programmes. These issues can be sorted out by adopting eco-friendly Biointensive Integrated Pest Management strategy.

Moreover, the use of synthetic chemicals has also been restricted because of their carcinogenicity, teratogenicity, high and acute residual toxicity, ability to create hormonal imbalance, spermatotoxicity, long degradation period and food residue (Dubey *et al.*, 2011; Khater, 2011).

Biointensive Integrated Pest Management (BIPM) is a systems approach to pest management based on an understanding of pest ecology. It begins with steps to accurately diagnose the nature and source of pest problems, and then relies on a range of preventive tactics and biological controls to keep pest populations within acceptable limits. Reduced-risk pesticides are used if other tactics have not been adequately effective, as a last resort, and with care to minimize risks (Benbrook, 1996). The most recent Biointensive integrated approaches for pest management utilizes components such as bio-control agents *viz.,* predators, parasitoids, and other entities such as Pheromones for monitoring and management of pests, Use of bio-rational chemicals like pheromones and other allelochemicals, entomopathogens such as mycorrhizal fungi, *Bacillus thuringiensis,* Nuclear Polyhedroviruses (NPV) and use of botanicals and Insect growth regulator (IGR) and genes used to transform crops to express resistance to pests (Copping and Menn 2000). The most advantage of Biointensive Integrated Pest Management is a cumulative effect, its specificity, eco-friendly, ease of adoption, compatibility with other components and eliminates or reduces use of pesticides. The benefits of implementing Biointensive Integrated Pest Management can include reduced chemical input costs, reduced environmental impacts, and more effective and sustainable pest management and such reductions will benefit the spice crops grower and in turn the society. Hence, Biointensive IPM is considered the desirable path for achieving sustainability in agriculture. The challenge before applied entomologists is to develop, validate and disseminate the site-specific Biointensive IPM technologies in management of insect pests in spice crops.

Insect Pests

I. Pepper (*Piper nigrum*)

1. Pollu Beetle, *Longitarsus nigripennis,* (Alticidae: Coleoptera)

Symptoms of damage

The damaged berries are seen with exit holes and later the berries dry up, turn dark and hollow and crumble when pressed. Irregular feeding holes on leaves. Grub may also eat the spike causing the entire region beyond it to dry up. When contents of one berry is exhausted, the grub move to next and feed continuously. Both grubs and adults are damaging stage (Regupathy and Ayyasamy, 2016).

Biology

Adult is a bluish yellow shining flea beetles. Eggs are laid on the berries and lay 1-2 eggs in each hole, egg period 5-8 days, larval period 30-32 days. Pupation occurs in soil in a depth of 5.0-7.5 cm. The pupal period is 6-7 days. Life cycle completes in 40-50 days. There are four overlapping generations in a year. The seasonal incidence of July and October are the peak period of infestation.

Management

Cultural operations like raking of soil and regulation of shade by standards. Tilling the soil at the base of the vine at regular intervals. Spraying of Neemgold (0.6 per cent) (neem-based insecticide) during August, September and October is effective for the management of the pest. The underside of leaves (where adults are generally seen) and spikes are to be sprayed thoroughly. Soil application of insecticides is also effective in controlling the grubs falling to ground for pupation. Spraying of vines with quinalphos (0.05%) twice a year during July and October control the pest effectively.

2. Top shoot borer, *Cydiahemidoxa* (Eucosmidae: *Lepidoptera*)

Symptoms of damage

The caterpillars damage terminal shoots by boring into them and drying of tender stalks and shoots. The affected berries appear larger. The incidence is more during August to December, when tender shoots are available.

Biology

The larvae are greyish green, 12-14 mm long, larval period is 10-15 days. Pupates inside shoots, pupal period is 8-10 days. Adult moth is tiny, forewing black with distal half red, hind wings are greyish. Life cycle completes in a month.

Management

Spraying vines with dimethoate or phosphamidon at 0.05% is effective. Use of parasitoids like *Apanteles* sp. (Braconidae), *Euderus* sp. (Eulophidae) and *Goniozus* sp. (Bethylidae) have been reported to attack the caterpillars in nature.

3. Gall forming thrips, *Liothrips (Gynaikothrips) karnyi* (Thripidae: *Thysanoptera*)

Symptoms of damage

It is a persistent pest in almost all the pepper growing areas of India. The thrips makes marginal galls on leaves within which they live in colonies and rasp, suck the sap and leaf tissue become thick. Under severe infestation, whole leaf gets crinkled or malformed with proliferation of cells results brittled leaves.

Biology

Eggs are laid in single within the marginal leaf folds or on the leaf surface, egg period lasts for 6-8 days. Nymphs whitish and sluggish, nymphal period is 9-13 days, pupal period, 2 to 3 days and adult longevity is 7-9 days.

Management

Spraying of vines with malathion (0.1%) or dimethoate (0.05%) or quinalphos (0.54%) is effective. An anthocorid bug and some predaceous mites have also been reported to be good biocontrol agents.Spraying of 2% neem oil and garlic emulsion will give better control.

II. Cardamom *(Elettaria cardamomum)*

1. Shoot, panicle and capsule borer, *Conogethes punctiferalis* (Crambidae: *Lepidoptera*)

Symptoms of damage

Early stage of the larva bores the unopened leaf buds and feeds on the leaf tissue. They also bore the panicles leading to drying up of the portion from the affected spot. Feed on immature capsules and the young seeds inside rendering the empty capsules. Late stage larvae bore the pseudostem and feed on the central core of the stem resulting in drying of the terminal leaf and thus produce characteristic 'dead heart' symptom. Oozing out of frass material at the point of tunneling is the indication for the presence of larva inside the plant parts. The incidence of this pest is noticed throughout the year but they occur in enormous number in four periods, December-January, March-April, May-June and September-October and their abundance synchronizes with the panicle production, fruit formation and new tiller production.

Biology

Eggs are pink, oval, flat and lays singly or in groups on the tender part of the plants. Larvae are long, pale greenish with a pinkish color dorsally, head and pro-thoracic shield brown in color and body covered with minute hairs.Pupation takes place in lose silken cocoon in larval tunnel. Adults are medium sized moth (22-24 mm) and the wings are pale in color.

Management

Parasitoids such as *Trichogramma chiloni*, *Tetrastichus* spp., *Apanteles* sp, *Eriborus* sp, *Friona* sp, *Braconid* wasp, *Telenomus* spp., *Xanthopimpla australicus*, *Agrypo*nsp etc. Predators *viz.*, Lacewing, ladybird beetle, King crow, dragonfly, spider, robber fly, reduviid bug, praying mantis, earwig, red ant etc.

2. Root grub, *Basilepta fulvicorne* (Eumolpidae: *Coleoptera*)

Symptoms of damage

The grubs feed on the roots in the form of irregular scraping. In advanced stage, the entire root system is found to damage and results in drying and rotting depending on the season of attack. In the severely infested plants, leaves turn yellow and dry.

Biology

Eggs are pale yellow in color and grubs are short, stout, 'C' shaped, pale white in color. The beetles are shiny, metallic blue, bluish green, greenish brown or brown in color.

Management

Neem cake @ 1 Kg/vine may be mixed with the soil at the time of planting.Predators: Lacewing, ladybird beetle, King crow, dragonfly, spider, robber fly, reduviid bug, praying mantis, earwig, red ant etc. will take care of grubs management.

3. Cardamom whitefly, *Dialeurodes cardamom* (Aleyrodidae: *Hemiptera*)

Symptoms of damage

Nymphs and adults found in colonies on lower surface of leaves. Leaves with chlorotic spots, turn yellow and dried up. Mainly damage to the plants is caused by the depletion of sap from leaves.In severe infestation the leaves turn yellow and the vigor, growth of the plant gets considerably reduced.The nymphs secrete sticky honeydew, which drops on to lower leaves. On these, black sooty mold develops, which interrupts photosynthesis of the leaves.

Biology

Eggs are cylindrical, pale yellow in color when freshly laid and gradually turn brown. Nymphs are elliptical and pale green in color. There are four nymphal stages. Adults are small soft bodied insect, about 2 mm long and having two pairs of white wings. The life cycle is completed within two-three week.

Management

Parasitoids, Encarsia spp., Eretmocerus spp., *Chrysocharis* spp. etc.Predators, Lacewing (*Mallada boninensis*), big-eyed bug, ladybird beetle, dragonfly, spider, predatory mites etc.Warm weather conditions are favorable for multiplication Natural enemies of whitefly.

4. Cardamom thrips, *Sciothrips cardamomi* (Thripidae: *Thysanoptera*)

Symptoms of damage

Infected panicles become stunted and shedding of flowers and immature capsules thus reducing the total number of capsules formed.Infestation causes formation of corky encrustations on capsule resulting in their malformed and shriveled condition.Such pods lack their fi ne aroma and the seeds within are also poorly developed.

Biology

Eggs are kidney shaped laid singly in the tender part of the leaf sheath, racemes, Nymphs tiny, slender, fragile and straw yellow in color, Adults are minute, dark greyish brown, 1.25 to 1.5 mm long and with fringed wings.

Management

Predators, Lacewings, big-eyed bug, *Oriuslaevigatus*, *Thripoctenus americensi* etc.

5. Cardamom aphid, *Pentalonia nigronervosa f. caladii* (Aphididae: *Hemiptera*)

Symptoms of damage

Both nymphs and adults suck up plant sap. Colonies of aphids are seen under concealed conditions inside leaf sheaths of the older pseudostems.

Biology

Nymphs are dark in color and adults are brown in color and have black veined wings. They reproduce by viviparous and parthenogenetically.

Management

Parasitoids,*Aphidius* spp., *Aphelinus* spp. etc, Predators: Ladybird beetle, lacewing, spiders, hover fly etc.

III. Turmeric and Ginger (*Curcuma longa & Zingiber officinale)*

1. Rhizome scale: *Aspidiotus hartii* (Diaspididae: Hemiptera)

Symptoms of damage

Both nymphs and adults infest rhizomes both in field and storage. The infested plants become weak, pale and withered in the field that results in shriveling of rhizomes and buds.

Biology

Scales are minute, circular; light brownish to grey with a thin pale membrane. It reproduces either ovovivparously or parthenogenetically Female lays about 100 oval, yellowish eggs under the scale. Egg period one day, nymphal period 30 days. Adult is yellow to deep yellow in color.

Management

Apply well rotten sheep manure / poultry manure in two splits @ 10 tons/ha, first before planting and the second at the time of earthling up. Spraying of neem oil @ 0.3 per cent or Neem gold @ 0.3 per cent or fish oil rosin @ 3% is also effective in controlling the pest infestation.Drench soil with dimethoate 30 EC or phosalone 35 EC @ 2 ml/L of water. Soak seed rhizomes, in insecticide solution of either dimethoate 30 EC or phosalone 1.5 ml/L or dichlorvos 0.5 ml/ L for 15 min. for storing.

2. Rhizome maggot: *Formosina flavipes* (Chloropidae: Diptera)

Symptoms of damage

Rhizomes and roots are tunneled extensively by the maggots resulting in rotting of rhizomes.

Biology

The egg incubation period is 3-4 days, while the duration of first, second and third larval instars is 2.00, 3.00 and 6.00 days respectively. The pupal period is 8.00 days. The longevity of adult fly 43-51 days.

Management

Avoid using seed material from the infested fields and spray dimethoate 500 ml in 500 -750 L water per ha. Soak seed rhizomes, in insecticide solution of either dimethoate 30 EC or dichlorvos 0.5 ml/L for 15 min. for storing.

3. Shoot borer: *Conogethes punctiferalis* (Pyraustidae: Lepidoptera)

Symptoms of damage

The caterpillar enters into the aerial stem killing the central shoot which results in the appearance of 'dead heart'.Yellowing and drying of leaves of infested pseudostems. The presence of a bore-hole on the pseudostem through which frass is extruded, withered and becomes yellowing of central shoot. The favorable temperature range of 30-33°C and relative humidity range 60-90% increases the pest infestation. The pest population is higher in the field during September-October and it is most active from July to October.

Biology

Female lays pale yellowish, oval flat eggs singly. The incubation, larval, prepupal, and pupal periods are 2.51, 13.25, 2.75 and 9.50 days respectively. The adult longevity of male and female was 7.50 – 9.00 and 8.00 days, respectively. It takes about 30.37 – 35.30 days with on an average of 30.65 days to complete life cycle from oviposition to adult emergence.

Management

Intercropping with cluster bean, cowpea, black gram, or groundnut (1: 2 ratio proportions) reduces shoot and capsule borer infestation and builds up natural enemies (*Microplitis, coccinellids*, spiders etc.) population. The infested shoots may be collected and destroyed.

4. Thrips, *Panchaetothrips indicus* (Thripidae: *Thysanoptera*)

Symptoms of damage

Thrips damage the undersides of leaves by sucking their plant sap. They damage young and soft parts of plants such as new leaves and shoots. Leaves become rolled up, and turn pale and gradually dry-up and in severe infestation they cause young leaves to wilt and dry out. The warm and humid conditions are favourable for their infestation.

Biology

Thrips reproduce by laying eggs and nymphs emerge from the eggs. It takes between 6-9 days to develop from eggs into adult thrips. Adult thrips are very small, have elongated abdomens and are yellowish or blackish in color. Although the adults have wings, these insect pests do not usually fly. They are often found on plants throughout all growth stages, from sprout development to tuber maturation.

Management

Spray of garlic chilli kerosene extract, Azadirachtin 10000 ppm+ *Lecanicillium lecanii* ($5X10^8$cfu/ml) and Chlorfenapyr 10 EC is effective in managing the thrips population.

5. Leaf roller, *Udaspes folus* (Hesperidae: Lepidoptera)

Symptoms of damage

Leaves become folded or rolled longitudinally. Complete defoliation takes place in severe condition. Presence of greenish larvae is seen inside the leaf roll. Temperature of 26-35°C and relative humidity ranging from 41-100% is congenial for the infestation.

Biology

The female normally lays a single egg on undersides of leaves. The egg is reddish, smooth and dome shaped. When about to hatch it turns white with a red top.The larva is greenish, sluggish & constructs its leaf shelter and comes out to feed only at night. It is smooth green with black head. Pupation takes place on the same plant within a cell. The pupa is long and cylindrical, watery green in color. The adult moths are brownish black. It has forewings with a white spots and hind wing with a large white patch, emerge in February or March and lay eggs before they die.

Management

Encourage predators such as spiders, parasitic wasps, predatory beetles, frogs and dragon flies. Use thorn wood on the leaves during hot weather to chase insects away. Apply insecticides such as alpha-cypermethrin, abamectin 2% or cartap hydrochloride to kill the larvae.

III. Coriander *(Coriandrum sativum)*

1. Coriander aphid, *Hyadaphis (Brevicoryne) coriandri* (Aphididae: *Hemiptera*)

Symptoms of damage

Infestation takes place on tender shoots and under surface of the leaves. There will be curling and crinkling of leaves with stunted growth with development of black sooty mould due to the excretion of honeydew.

Biology

Eggs are very tiny, shiny-black, and are found in the crevices of bud and stems of the plant. Aphids usually do not lay eggs in warmer parts of the world where viviparity is observed. Nymphs (immature stages) are young aphids; they look like the wingless adults but are smaller. They become adults within 7 to 10 days. Adults are small, 1 to 4 mm long, soft-bodied insects with two long antennae that resemble horns. Most aphids have two short cornicles (horns) towards the rear of the body.

Management

Natural enemies, *Aphidius colemani* and *Diaeretiella* sp., Fire ant, Robber fly, Big-eyed bug (Geocoris sp), Earwig, Ground beetle, Cecidomyiid fly, Lacewing, Ladybird beetle, Spider, Preying Mantid, Reduviid, Dragon fly, hoverfly, etc.

2. Cutworm, *Spodopteralitura* (Noctuidae: *Lepidoptera*)

Symptoms of damage

In early stages, the caterpillars are gregarious and scrape the chlorophyll content of leaf lamina giving it a papery white appearance.Irregular holes are produced on leaves initially and later skeletonisation occurs, leaving only veins and petioles.

Biology

Female lays about 300 eggs in clusters. The eggs are covered over by brown hairs and they hatch in about 3-5 days. The caterpillar measures 35-40 mm in length, when full grown. It is velvety, black with yellowish – green dorsal stripes and lateral white bands with incomplete ring – like dark band on anterior and posterior end of the body. It passes through 6 instars. Larval stage lasts 15-30 days. Pupal stage lasts for 7-15 days and pupation takes place inside the soil. Moth is medium sized and stout bodied, forewings with pale grey to dark brown in color having wavy white crisscross markings. Hind wings are whitish with brown patches. Moths are active at night. Adults live for 7-10 days. Total life cycle is completed in 32-60 days depending on environmental conditions.

Management

Parasitoids: Trichogramma chilonis, Telenomus spp., Campoletis chloridae, Peribeaor bata, Glipapanteles africanus, Cotesia ruficrus, Chelonus carbonator, Blepherella setigera, Bracon spp., Sarcophaga dux, Sarcophaga albiceps, Brachimeria lasus, Lasiochalcidia erythropoda.

Predators: Chrysoperlazastrowisillemi, C. crassinervis, King crow, Braconid wasp, Dragon fly, Spider, Reduviid, Preying mantid, Harpactor costalis, Rhynocoris fuscipes, R. squalis, Polistes stigma, Coranus spiniscutis, Andrellus spinidens, etc.

3. Mite, *Tetranychus telarius Linn. (*Tetranychidae*: Acarina)*

Symptoms of damage

The mites frequently attack the coriander crop and whole plant becomes whitish yellow and appears sickly. It mostly feeds on young leaves and the infestation is more severe on young inflorescence. Mites are seen on the lower side of the leaves and when serious, cause webbing and feed from within the web. Plants get stunted at severe infestation.

Biology

Eggs are microscopic, hyaline, and globular, laid in masses beneath clods and are either active (red in color) or dormant (white in color). Nymphs are yellowish in color and the adult mites are very small measuring about 0.5 mm in length, metallic brown to black in color. Their forelegs are distinctively longer than the other three pairs.

Management

Predators: *Anthocorid bugs (Orius spp.), mirid bugs, syrphid/hover flies, green lacewings (Mallada basalis and Chrysoperla zastrowisillemi.), predatory mites (Amblyseius alstoniae, A. womersleyi, A. fallacies and Phytoseiulus persimilis), predatory coccinellids (Stethorus punctillum), staphylinid beetle (Oligota spp.), predatory cecidomyiid fly (Anthrocnoda xoccidentalis), predatory gall midge (Feltiella minuta), Predatory thrips etc.*

4. Thrips, *Thrips tabaci* (Thysanoptera: *Thripidae*)

Symptoms of damage

Thrips have a very peculiar feeding behavior. They start the feeding by rasping and sucking and they lacerate the leaf surface with their mouth parts to release the liquids from the plant cells. In this process, thrips release chemicals substances that help to predigest the cell sap. The damaged leaves show silvery patches or streaks that shine in the sun. When damage is severe, these small patches can occupy most of the surface of the leaf and the plant cannot adequately photosynthesize. The plant loses more water than normal through the damaged tissues which are the entry points for plant pathogens.

Biology

Eggs are microscopic and white or yellow in color which is inserted one by one in the plant tissues by the females. Only one end of the egg will be near the surface of the tissue to allow the immature to emerge. Adults prefer to lay their eggs in leaf, cotyledon or flower tissues.

Nymphs are very small, pale yellow to brown in color and the pupae appear as an intermediate form between the immature and the adult. Nymphs have short antennae and the wing buds are visible but short and not functional. They are found at the base of the plant neck or in the soil.

Adults are up to 2 mm in size, pale yellow to dark brown in color and have fully developed wings which are very different from other insects. They have a single longitudinal vein in which there is several hairs connected perpendicular to the vein. The wing appears as fringe with hairs. When at rest, the wings are folded along the back of the insect. Besides undersurface of leaves, they can also be found in the flowers. Adults are more mobile than immature and are attracted to yellow and white colors.

Management

Parasitoid: *Ceranisus menes;* Predators: Predatory mite, Predatory thrips, *Oligota* spp., *Orius* spp. (pirate bug), Hover fly, Mirid bug, etc.

IV. Onion and Garlic, *(Allium sativum)*

1. Onion thrips, *Thripsy tabaci (*Thripidae*: Thysanoptera)*

Symptoms of damage

Both adults and nymphs feed within the mesophyll layer using a punch-and suck motion. The beak and mandible is thrust forward to puncture the leaf epidermis and sap released from injured plant cells is sucked up.Removal of chlorophyll causes the feeding area to appear white to silvery in color.Areas of leaf injury can occur as patches and streaks.When feeding injury is severe, leaves take on a silvery cast and can wither.Tiny black "tar" spots of thrips excrement are evident on leaves with heavy feeding injury.

Biology

Eggs are white to yellow, kidney-bean shaped, microscopic in size. Develop within leaf tissue with one end near the leaf surface. Egg duration is 5-10 days. Nymphs are of Instars I and II are active, feeding stages. White to pale yellow, elongate and slender body. Resemble adult, but without wings. Antennae are short and eyes are dark in color. Feed on new leaves in the center of the garlic neck. Crawl quickly when disturbed. Larval duration is 10-14 days.

Adult: About 1.5 mm long; elongate, yellow and brown body with two pairs of fringed (hairy) wings. Parthenogenic (asexually reproducing) females; males are extremely rare. Feed on young leaves in center of garlic neck and insert eggs individually into leaves. Fly readily when disturbed. Adult life span is about 1 month.

Management

Blue pan water/sticky traps 15 cm above the canopy for monitoring thrips @ 4-5 traps/acre. Locally available empty tins can be painted blue and coated with grease/Vaseline/castor oil on outer surface may also be used. The Light traps traps @ 1 trap/acre, 15 cm above the crop canopy for monitoring and mass trapping of insects. Light traps with exit option for natural enemies of smaller size should be installed and operate around the dusk time (6 pm to 10 pm). Parasitoid: *Ceranisus menes* (nymph) Predators: Syrphidflies, minute pirate bug/ anthocorid bug (*Blaptosthethus* sp, *Buchananiella white*, *Orius tantilus*), praying mantis, predatory thrips (*Aeolothrips fasciatum*), damsel bug, lace wings, *Coccinellids* (*Cheilomenes sexmaculata*), spiders etc.

2. Cutworm, *Spodoptera litura* (Noctuidae: *Lepidoptera*)

Symptoms of damage

In early stages, the caterpillars are gregarious and scrape the chlorophyll content of leaf lamina giving it a papery white appearance. Irregular holes are produced on leaves initially and later skeletonisation occurs.

Biology

Female lays about 300 eggs in clusters. The eggs are covered over by brown hairs and they hatch in about 3-5 days. The caterpillar measures 35-40 mm in length, when full grown. It is velvety, black with yellowish-green dorsal stripes and lateral white bands with incomplete ring-like dark band on anterior and posterior end of the body. It passes through 6 instars. Larval stage lasts 15-30 days. Pupal stage lasts for 7-15 days and pupation takes place inside the soil. Moth is medium sized and stout bodied, forewings with pale grey to dark brown in color having wavy white crisscross markings. Hind wings are whitish with brown patches. Moths are active at night. Adults live for 7-10 days. Total life cycle is completed in 32-60 days depending on environmental conditions.

Management

Parasitoids: *Trichogramma chilonis, Telenomus spp., Campoletis chloridae, Peribea orbata, Glipapanteles africanus, Cotesia ruficrus, Chelonus carbonator, Blepherella setigera, Bracon spp., Sarcophaga dux, Sarcophaga albiceps, Brachimeria lasus, Lasiochalcidia erythropoda.*

Predators: *Chrysoperlazastrowisillemi, C. crassinervis, King crow, Braconid wasp, Dragon fly, Spider, Reduviid, Preying mantid, Harpactorcostalis, Rhynocoris fuscipes, R. squalis, Polistes stigma, Coranus spiniscutis, Andrellus spinidens, etc.*

3. Onion Maggot, *Deliaantiqua* (Anthomyiidae: *Diptera*)

Symptoms of damage

Damage from the first generation is seen in mid-to-late June as wilted onion seedlings. If you try to pull the wilted plant, it is likely to break just below the rotting stem of the seedling. Quite often the seedling dies before the maggot is full-grown. When this occurs the maggot will move down the row to the next seedling. The damage appears as dying patches in a row rather than as single plants. Damage can also be determined by pulling the plants and observing the maggot infestation. Plants damaged at this time will result in grossly misshapen onions that are not suitable for market. In addition, secondary rots may set in following the onion maggot damage and the bulb will be totally destroyed.

Biology

The onion maggot passes the winter in the soil in a dormant stage. Pupae are brownish in color, oval and slightly larger than a grain of wheat. As soon as the weather warms up in spring, the pupae develop into adults which emerge in mid-to-late-May in the onion-growing areas. These adults are grayish and look something like the common housefly except that the onion maggot has larger legs and a narrower abdomen.

Management

Avoid close spacing while planting and follow crop rotation with precaution to keep field under sanitation. Conserve predators such as ground beetle, rove beetles, spiders etc. by providing grassy refuge strips.Limited natural control of the onion maggot is provided by a complex of predators and parasitoids which include the predatory flies *Coenosiatigrina* and *Scatophagastercoraria*, the wasp, *Aphaerata pallipes*, the beetle *Aleochara bilineata* and a fungus *Entomopthora muscae*.

4. Leaf Miner, *Liriomyza* spp.

Symptoms of damage: Damage caused by leafminers is primarily cosmetic; however, contamination by pupae and larvae is a marketing problem for green bunching onions. Damage in dry onions and garlic is rarely a concern unless leafminers become so numerous that they prematurely kill foliage.The potential for damage to onion and garlic will be higher when other hosts are grown nearby.

Biology

The developmental thresholds for eggs, larvae, and pupae are at 9 to 12°C. The combined development time required by the egg and larval stages is about seven to nine days at warm temperatures (25 to 30°C). Another seven to nine days is required for pupal development at these temperatures. Both egg-larval and pupal development times lengthen to about 25 days at 15°C. At optimal temperatures (30°C), the leaf miner completes development from the egg to adult stage in about 15 days.

Management

Thoroughly check fields of previously planted with susceptible crops before planting onion. Allow at least two weeks for leafminer flies to emerge from pupae in the soil before planting onion in a field that previously had a leafminer problem. Biological control by naturally occurring predators and parasites is often effective in controlling this pest in organically grown onion and garlic crops. However, supplemental releases of commercially available natural enemies are rarely economically viable. Natural enemies, especially parasitic wasps, can reduce leafminer numbers. However, these wasps are very susceptible to insecticide sprays, so they may not be reliable or effective in fields where insecticides have been used. Cultural controls as described above are critical. Azadirachtin products and the Entrust formulation of spinosad can also be used.

References

Benbrook, C.M. (1996). Pest management at the cross roads (272 pages). Consumers Union, Yonkers.

Copping, L.G. and Menn, J.J. (2000). Biopesticides: A Review of Their Action, Applications and Efficacy. *Pest Management Science*, 56, 651-676.http://dx.doi.org/10.1002/1526-4998.

Devasahayam S, Premkumar T, Koya KMA (1988). Insect pests of black Pepper Piper nigrum L. in India – a review. *Journal of Plants and Crops* 16:1-14.

Dhanya M.K., Rini C.R., Ashok kumar K., Murugan M., Surya R. and Sathyan T (2019). Biointensive approaches for management of pests and diseases in small cardamom and black pepper. In: Biointensive approaches: application and effectiveness in the management of plant nematodes, insects and weeds. Editors: M.R. Khan, A.N. Mukhopadhyay, R.N. Pandey, M.P.Thakur, Dinesh Singh,M.A. Siddiqui, Md. Akram, F.A. Mohiddin and Z. Haque. Today & Tomorrow's Printers and Publishers, New Delhi-110002, India, Pp. 549-585.

Dubey, N.K., R. Shukla, A. Kumar, P. Singh and B. Prakash, (2011). Global Scenario on the application of Natural products in integrated pest management. In: *Natural Products in Plant Pest Management*, Dubey, N.K., CABI, Oxfordshire, UK., pp: 1-20.

Khater, H.F., 2011. Ecosmart Biorational Insecticides: Alternative Insect Control Strategies. In: Insecticides, Perveen, F. (Ed.). In Tech, Rijeka, Croatia, ISBN 979-953-307-667-5.

Prakash, A., J. Rao and K. Revathi, (2016). Recent Advances in Life Sciences, Proc. XV AZRA International Conference, at Ethiraj College for Women, Chennai, AZRA, India, Pp 250.

Regupathy A., and Ayyasamy, R (2016). A Guide on Crop Pests. Sixth Revised and Enlarged Edition, 2016. Pp.389.

For Product Safety Concerns and Information please contact our EU
representative GPSR@taylorandfrancis.com
Taylor & Francis Verlag GmbH, Kaufingerstraße 24, 80331 München, Germany